D0793785

the BLACKWINGED NIGHT

the BLACKWINGED NIGHT

Creativity in Nature and Mind

F. DAVID PEAT

HELIX BOOKS

PERSEUS PUBLISHING
Cambridge, Massachusetts

Many of the designations used by manufacturers and sellers to distinguish their products are claimed as trademarks. Where those designations appear in this book and
Perseus Publishing was aware of a trademark claim, the designations have been
printed in initial capital letters.

Copyright © 2000 by F. David Peat

All rights reserved. No part of this publication may be reproduced, stored in a retrieval
system, or transmitted, in any form or by any means, electronic, mechanical, photo-
copying, recording, or otherwise, without the prior written permission of the pub-
lisher. Printed in the United States of America.

A CIP catalog record for this book is available from the Library of Congress.
ISBN 0-7382-0205-3

Perseus Publishing is a member of the Perseus Books Group.

Find us on the World Wide Web at http://www.perseuspublishing.com

Perseus Publishing books are available at special discounts for bulk purchases in the
U.S. by corporations, institutions, and other organizations. For more information,
please contact the Special Markets Department at HarperCollins Publishers, 10 East
53rd Street, New York, NY 10022, or call 1-212-207-7528.

Text design by Jeff Williams
Set in 14-point New Aster by the Perseus Books Group

First printing, March 2000

1 2 3 4 5 6 7 8 9 10—03 02 01 00

For James

Contents

chapter 1

Introduction

We are all creative. Each one of us. We are creative when we're asleep and dreaming. We are creative the moment we open our eyes and look at the world. We are creative when we speak. Indeed, we have been immersed in creativity from the moment of birth. Creativity nurtured us when we took our first steps and spoke our first words. And creativity will continue to embrace us until the end of our lives.

Creativity permeates the cosmos. It is the driving force that sustains the elementary particles, the stars and galaxies, and even time itself. Creativity is the stuff of our bones. It surges though the body with each beat of the heart. It is that compelling power that directs salmon from the sea on hazardous journeys to spawn upriver. It is the force that drives the plant and flower. It is so much a part of us that we can be forgiven for sometimes forgetting its constant existence.

We all have access to creativity. At times, we summon it to use it in our work and daily lives. But at other times, creativity is a blind, amoral power that makes use of us as its vehicle. Creativity can arrive in a dream. Or it may result from a long struggle. It can appear as a sudden, dramatic insight or as the product of the years of hard work needed to perfect a scientific theory, a novel, or a piece of sculpture. One thing is certain: Creativity is ever present. It is a force to be enjoyed or endured but above all celebrated. Creativity is free, alive, and spontaneous.

We cannot control creativity or bend it to our will. Neither can we trap it within the confines of a definition. But for the purposes of this book, we need to uncover its disguises. For this reason, we shall look at creativity from three different angles—but always remembering that the boundaries of this approach are loose and that creativity itself escapes every definition.

Let us therefore consider creativity in three of its aspects:

1. Making something new, original, or unexpected
2. Renewing and sustaining what already exists
3. Healing and making things whole

Making It New

"Make it new" was the maxim of the poet Ezra Pound and the whole modernist movement. In the early part of the twentieth century, people were both shocked and excited by the ever changing movements in the arts. It was the Jazz Age—the time of flappers, Charlie Chaplin, the Charleston, and the Black Bottom. It was the era of F. Scott Fitzgerald's Lost Generation, expatriates in Paris, Ernest Hemingway's short stories, and Amelia Earhart's flight around the globe. Women had achieved the right to vote. People were decorating their homes in an entirely new range of colors and fabrics. There were new dances to learn, new music to listen to, new fashions to wear, new clubs to visit, and new cocktails to drink. Cole Porter celebrated some of those icons and innovations in "You're the Top": Mickey Mouse, a Waldorf salad, the hands on a Roxy usher, the feet of Fred Astaire, Garbo's salary, a Coolidge dollar, Mrs. Astor, an Arrow collar, Pepsodent, and cellophane.

During the early years of the twentieth century, being creative meant producing something totally different, something never seen before, and, if possible, something that would shock the public in the bargain. In the Cabaret Voltaire in Zurich, Tristan Tzara, Emmy Hennings, and Hugo Ball overwhelmed the public with their readings and the events of the dada movement. Audiences caused a near riot in Paris at Stravinsky's *Rite of Spring*. In fact, it was said that Debussy had tears in his eyes at the first performance because he realized he had been surpassed by the new. Then, a few years later, even Stravinsky found himself behind the times as Schoenberg's serialism in music became the mark of the avant-garde.

Change was everywhere. In 1899, Guglielmo Marconi successfully broadcast radio signals between England and France. In Paris, Alexandre-Gustave Eiffel's 984-foot tower was already ten years old. By 1909, Louis Bleriot was crossing the English Channel in his monoplane while Gertrude Stein was pushing language past its former limits with her writing. A few years later, in Trieste, James Joyce brought out sections of *Finnegans Wake* as *Work in Progress,* and from a London bank, T. S. Eliot was working on *The Waste Land.* In Italy, Filippo Tommaso Marianetti praised the radically new, fast, and powerful and poured scorn on the old in his *Futurist Manifestos*. From Spain came Dali and Picasso. Even dada was soon out of date, giving way to surrealism. Cinema was on the move with the experiments of Cocteau, Dali, Brunel, Eisenstein, Vertof, and a host of others. Antonin Artaud proposed his theater of cruelty, and, before his mental breakdown, Vaslav Nijinsky was creating a new choreography.

Not only the arts but also science was reeling from a series of revolutions. Scientists had thrown out two hundred years of Newtonian physics and replaced it with Einstein's relativity. Quantum theory abandoned certainty and embraced Heisenberg's indeterminacy. The certitude of the material world dissolved into a flux of atoms, elementary particles, and collapsing wave functions. Freud revealed his theories of the unconscious and infant sexuality. In mathematics, Kurt Gödel would eventually show that all great mathematical systems are inherently incomplete and that such logical systems can never be reduced to a finite series of fundamental axioms. Finally, by the middle of the twentieth century, biologists had discovered the key to genetic inheritance and were beginning to decode the DNA molecule.[1]

Chaos theory revealed that rather than the universe being governed by Newtonian clockwork, randomness and chaos lie at the heart of all of nature's systems. In quantum theory, John Bell showed that our notions of space and separation simply do not apply in the subatomic world; pairs of initially correlated elementary particles continue to remain correlated even when separated by very long distances and in ways that transcend commonsense notions of space and time. Add to all this change the impact of new technology and scientific instruments—radio telescopes, electron microscopes, space probes, high-speed computers, television, jet aircraft, modern pharmaceuticals, organ transplants, global communications, the Internet, and even the possibility of some form of artificial intelligence (AI).

Other centers of change rose to prominence during the twentieth century. In the 1960s, as England emerged from its austere postwar years, its once staid capital became known as "swinging London." For a short period, everything seemed possible as a host of new musicians, actors, photographers, fashion designers, and filmmakers dominated mass culture. Then from Haight-Ashbury came the period of love, peace, and psychedelics, which promised to turn us on and transform the world. But, like the 1968 student riots in Paris, these two periods of creativity petered out to leave us only warm, nostalgic memories (and a stack of treasured LPs).

The 1960s were an exciting time, but in terms of generating the new, the first decades of the twentieth century were even more innovative. Ideas changed in a far more radical way and at a higher speed than the human race had ever before experienced. Today, nearly a century after relativity and the quantum theory were introduced, we are still struggling to grasp

---

[1]Indeed, the revolutions in science were not confined to the first years of the twentieth century but continue to the present day with the search for ultimate theories and origins. Only a few decades ago, the science of chaos theory was born, and its implications are still being explored.

these new concepts and translate them into more human terms.

Inevitably, the impact of such high-speed change gave rise to the notion that the only way to be truly creative was to be different. Novelty became the new aesthetic. It was the yardstick against which creativity, in the early decades of the twentieth century, was measured. Woe betide any artist, writer, or musician whose style did not adapt to the times or who began to revisit earlier work. Such an artist was considered played out as a creative force.

Think of Picasso and Stravinsky, the two icons of this modern attitude toward constant novelty and change. There may have been other artists and composers equally as talented, yet when it comes to innovation, these two cannot be matched. Indeed, unlike other creators, they cannot be associated with any one style. Stravinsky's compositions move through a number of periods, from early works such as the lyrical *Fire Bird*, through the dissonant, rhythmic shocks of *The Rite of Spring*, into the neoclassical *Pulcinella* and *Soldier's Tale,* and on to the *Symphony of Psalms* and *Oedipus Rex* or concise and heavily concentrated serialist pieces such as *In Memoriam Dylan Thomas*.

Picasso was similarly innovative in a wide variety of styles and media. After he had moved through his Blue and Rose Periods and the shocking *Les Demoiselles d'Avignon,* he invented cubism (along with Georges Braque) and then moved on to neoclassicism. From Picasso too came the idea of using ordinary materials to make sculptures—a monkey created out of a toy car, a bull from a bicycle seat and handlebars—a move that revolutionized the field. Films of Picasso in old age show his elfin nature, as he picked up the fish bones from his lunch and pressed them into clay to create a new design for a plate. His genius lay in his childlike ability to play and always see the world in a fresh way and then translate this into his art.

Recent biographers of Picasso have dwelled on his treatment of women and his love of money, which only emphasizes an earlier point about creativity: At times, it can be a blind and amoral force that chooses the vehicle though which it enters.

Picasso may have behaved unthinkingly and even brutally in his personal life, but he had made a Faustian pact with the creativity within him. In this, he conforms to the heroic image of the creator. And this exposes another issue of this book—the degree to which we have extolled the heroic artist's struggle against all odds but have forgotten the other ways in which creativity manifests itself.

Modernism or Postmodernism: Questioning the New

It seems that there are no Picassos or Stravinskys around today, and the sense of an accelerating pace of change, with "the new" appearing each year in music, art, writing, and science, is long gone, but the assumption that there is a necessary connection between creativity and change remains. In the pages ahead, I am going to challenge that assumption, along with the ethos of the hero-creator who believes that being different is the only way of being creative.

Innovation continues in the sciences, and schools of art are constantly changing, but none of this has the same dramatic impact as what went before. In fact, modernism has been replaced by postmodernism (or perhaps the oceanic waters of deconstruction have simply dissolved it). Nevertheless, the ethos of the new persists. In fact, it is one of the presiding values of our age, together with progress, efficiency, and accumulation of everything from money and goods to facts and knowledge.

Some cultural critics have questioned the ethos of novelty and change and the maxim that "more is better." In recent years, we've become concerned with nature and the environment. We feel guilty about our overconsumption and our excessive use of energy. We question the dream of unlimited progress and talk about sustainability in place of continuous change.

Despite that, new models of cars are exhibited at the automobile shows each year, and some people still feel that they must change models every year or two to demonstrate their success and achievement. Similarly, designers are in desperate competition to bring out the latest spring fashions. Pharmaceuticals have to be new and better. It's no longer enough to invite your friends for a meal; now, you must offer them whatever cuisine is currently fashionable (sushi, Thai, tapas, you name it—who knows what it will be by the time this book comes out!). What's more, the decor of a restaurant definitely has to be "in." One month, our children are being mugged for wearing the latest running shoes; the next, the mugger wouldn't be seen dead in them. Even the PC on which I'm typing is going to be several models out of date by the time this book is finished. And so, despite the warnings of environmentalists, the ethos of the new remains firmly entrenched.

In the first decades of the twentieth century, the new was genuinely intoxicating. Today, it's merely addictive. We hardly notice that we can't do without speed, novelty, and the unexpected. Television programs made for young children are constructed with scenes that change every minute or so because experts believe that a child's attention span is so short that novelty must always be present. As a result, we condition a new generation of television viewers to be ever more accepting of change. It is even said that computers and computer games are speeding up children's nervous systems so that the demand for change is going to increase.

There is nothing inherently wrong in celebrating the new or enjoying the excitement we experience when encountering the unexpected. After all, what is more engaging than watching a baby take its first steps or speak its first words? The problem arises when things get out of balance and we confuse simple novelty with creativity and stasis with its absence. But the high value we currently place on novelty is actually something of an aberration. The history of Western civilization and of so many of the world's other cultures tells a very different story.

The Role of Tradition

Not everyone welcomes unlimited change. Indigenous and peasant cultures, for example, emphasize doing things in traditional ways. To our contemporary attitudes, members of these cultures may seem to be overly conservative sticks-in-the-mud. But from their own perspective, time-honored approaches to farming, fishing, and hunting as well as their family structures, marriages, and religious ceremonies have worked effectively and held society together for centuries and maybe even millennia—often in harmony with the environment. That is not to say that new and creative innovations do not occur in such cultures. Certainly, they do. Peoples change, populations migrate, climates alter, and new methods are sometimes introduced. But, in keeping with their worldviews, people in these societies often associate change with some sanction from a higher authority. In other words, change must be absorbed into the historical creation stories that cement society together and provide its cohesion and collective sense of meaning.

In ancient China, the turtle oracle (precursor to the *I Ching*) was given to the Shang when a turtle with special markings on its back climbed out of the Yellow River, and writing and cooking were associated with historical figures that were partly divine. From Greece, Plato related that the Egyptian god Theuth invented writing, geometry, and astronomy. The great oceangoing canoe of the Haida people from the Pacific West Coast—an impressive piece of marine engineering—was a gift from Raven. Buffalo Calf Woman gave the pipe to the Plains Indians, and it became the center of all their sacred ceremonies. In each case, the new was praised not so much for its difference and innovation but for the way it integrated into tradition.

Tradition is given such high value because with innovation, there is always the danger that the unexpected will disrupt the stability of society in unexpected ways. In our own time, for example, take the development of new seed stocks in the industrial world, designed to produce higher yields per acre. At first sight, this advance would appear to be a blessing to Third World countries whose populations lack protein. But the importation of seeds into a traditional culture also means that that society could become locked into the industrial world's chemical industry, for some of these crops are ecologically fragile and require accompanying fertilizers, pesticides, herbicides, and even ripeners. In turn, traditional farming methods simply may not work for such intensive farming. And so, in a short space of time, a traditional agrarian society can be dramatically disrupted.[2]

Traditional societies are cautious about the new, for the ramifications of the unexpected cannot be predicted or controlled. When making a decision, the elders of the Iroquois people do not consider what a given change will mean for their own lives but how it will impact the seventh generation to come after them. Contrast that with the limited vision of so many of our own leaders, whose major concerns are staying in office and getting reelected in a few years' time.

Traditional peoples realize that knowledge is a powerful force that must be treated with respect and integrated into the fabric of society. Therefore, before individuals in a traditional society can practice a new task or ceremony, they must be properly sanctioned. This means that they must be given permission by whoever has a special relationship with that task or ceremony—a power, a spirit, a keeper of the animals, or a member of another group who has already been sanctioned. This is equally true in regard to the way young people in such a society learn new things as they grow. Learning is accomplished by carefully watching the way adults carry out their tasks, internalizing this knowledge, practicing in private, and, finally, performing the new task in public to obtain formal acknowledgment that the young person now has an official relationship with the new knowledge.

This respect for tradition and the past is not the province of indigenous peoples or small farming societies alone. Consider the great explosion of creativity in art, music, and literature that began in fifteenth-century Florence. This creative burst

---

[2]One need not go to the Third World for such examples. In other parts of the world, for instance, the drop in infant mortality, produced by better hygiene and diet, rapidly led to overpopulation. And who would have predicted that the invention of the birth control pill would have produced the sexual revolution of the 1960s? Even as this book is being written, the long-term impacts of genetically modified foods on the environment and the human immune system are being hotly debated.

was not called "the *new* birth" (or the *new* anything else, for that matter) but rather *la rinascita*—a *re*-birth or Re-naissance.

The entire ethos of the Renaissance was not about striving to produce something new and totally original but about the return to a glorious past—the world of Athens and the great Greek philosophers Aristotle and Plato, as well as the Greek and Roman sculptors whose work was being unearthed at that time. However, it was not a return in the sense of a mere copying or lifeless imitation but as an *imitato*—a grasping of the genius of the past, a true revival, and a reanimation. It was a return to the forms of the past and also to their artistic and humanistic spirit. Thus, the fourteenth-century poet Petrarch called for a reawakening from the "slumber of forgetfulness" into "the pure radiance of the past" (Giorgio Vasari, *Lives of the Great Artists*, 1550).

When Marsilio Ficino created his Florentine Academy, under the sponsorship of the Medici, he modeled it upon Plato's academy and invited the leading artists and thinkers of Florence to meet and discuss the aesthetics of the Golden Age and plan works that would arise out of that particular ambiance.

Another Florentine, Giorgio Vasari (who named the age *la rinascita*), wrote about these artists and their immediate predecessors in his *Lives*. While praising them for their skill and accomplishments, he often reserved his greatest acclamation for those who had managed to capture something of the glory of the ancient style. After all, if the ancient Greeks had perfected the depiction of the idealized human body, why should anyone wish to devise a radically different approach? Thus, the painter Giotto was praised as "Giotto alone who, by God's favor, rescued and restored the art," rather than as Giotto the great innovator who wished to change art (*Lives of the Great Artists*, 1550).

Of course, artists such as Giotto and Masaccio were considerable innovators. Vasari praised the latter, who introduced perspective into painting, for originating a new style. But it was new in the sense of correcting the crude, clumsy, and artificial style of the past, that past that represented a fall from the triumphs of Greece and ancient Rome. Masaccio was part of the rebirth, for his work was all that was living, realistic, and natural.

Before the Renaissance, artists had worked within a traditional ethos. The great center of art before Florence had been Siena, where painters such as Duccio, Simone Martini, and the Lorenzetti brothers worked within particular stylistic templates. In rendering a crucifixion or the birth of the Virgin, facial expressions and gestures were fixed by convention, as were the particular colors of garments. In the same way, poets wrote within the sonnet template invented by Petrarch, which continued to be used (albeit in modified English form) by Shakespeare, Milton, and Wordsworth.

In Japan, actors of the Kabuki and No theaters still spend a long apprenticeship to learn the traditions of voice, movement, and gesture. Until the Cultural Revolution, Chinese scrolls were painted according to tradition. Icon painters in the Russian and Greek Orthodox traditions used pattern books to work from a prototype. *The Vladimir Virgin*, for example, supposedly painted by Saint Luke and brought to Kiev in the twelfth century, was the generative source of many icons all over Russia.

In music, Gregorian chants were sung in the tradition of particular monasteries. Centuries later, Bach was content to create his major works within the confines of the fugue, a form that was also adopted by Mozart and Beethoven and even, in our own time, by Stravinsky and Bartók.

Renewal

And so, we have arrived at our second use of the word *creativity*. Creativity does not always mean the production of the new and different; it can also mean an act of renewal and revitalization. There is a world of difference between a Florentine sculptor mechanically copying a Greek statue and Michelangelo creating his *David*. The latter is both an act of deep respect for the past and an artistic feat involving the reanimation of an older form. (An amusing story is told of Michelangelo's youth. While working in Florence, he carved a life-sized cupid with such skill that it could pass for the work of a sculptor from ancient Rome. Some said that Michelangelo artificially aged the piece. Others said that one Baldassare del Milanese buried it in his garden and then sold it as a genuine antique. In any event, the artist's contemporaries believed this piece to be a genuine antique for several years. The final discovery of its originator enhanced Michelangelo's reputation, rather than denigrating him as a mere copier or forger.)

Take another example, this time from poetry. The sonnet, developed in the Italy of Dante and Petrarch, was a perfect form for Tuscan poetic expression, but it did not adapt well to the particular rhythms of the English tongue. The Earl of Surrey performed a creative act of transformation in the sonnet, which was, in turn, continued by Shakespeare, Spencer, Milton, and Wordsworth. The change was not gratuitous, for it was needed to satisfy the demands of a different language and mode of speech. Thus, it was not so much about producing the new for novelty's sake alone; rather, it was an act of creative transformation that arose out of respect for a particular poetic form.

For centuries, poetic language in Italy itself remained literary to the point of being associated with a particular social class and political attitude. Then, in the 1940s, a writer from the political Left, Pier Paolo Pasolini, took for his subject the life of the streets of Rome. In doing so, he created a new "poetry of the city" (*poesia civile*). As happened with the transformation of the sonnet form in English, a poet of genius paid homage to a traditional form while, at the same time, performing a creative act to reanimate and transform. Even an icon painter working from a book of patterns is still creating something fresh and vibrant that arises out of a deep respect for the prototype.

I have often told my friends the story of David the baker. When I lived in Ottawa, David had a bakery a few blocks from my house, and I'd walk there each morning to buy a loaf of wholewheat bread. There was always an exciting smell coming from his bakery, and I once remarked to him that, though all the loaves looked the same and weighed the same, each day when I ate the bread, it tasted so fresh. "That's it," he said. "That is how I exercise my creativity. My creativity is to make bread every morning as best as I can."

David the baker was using exactly the same creativity that Michelangelo put into his *David* or Edmund Spenser put into the sonnet. An identical creativity is present when we are feeling good about ourselves. Each morning seems fresh, new, and full of potential, yet it is nonetheless part of the one cycle that continues forever. At moments like these, we should recognize just how creative we are, for it is we who are painting the morning in its glorious colors, we who are creating the new morning, we who are merging our horizons with the glory and the freshness of the world.

Years ago, when I was living in London, I had a season ticket for the Proms. One evening, the concert featured the Tchaikovsky Piano Concerto no. 1, and I wondered if I really wanted to go out that night to hear such an old warhorse. As it turned out, the concert was a remarkable experience. The pianist in question happened to be Radu Lupu at the start of his career, and the music I heard was totally original. I felt I was hearing Tchaikovsky for the very first time. That familiar work sounded new to me. Each phrase was unexpected and totally engaging. It was a performance of considerable creativity from a young pianist who had taken something so well known and re-created it in an entirely new way.

To say what was new about it would have been extremely difficult for a layman like myself, for the notes were played as written. The creativity lay in such subtle details as the micro intervals of time when notes were delayed or anticipated, the attack and emphasis of a particular note, the overall flow of a phrase, subtle changes of dynamics, and so on. It has been said that "God lies in the details," and so can the creativity of renewal. Something enormously original can reside within an almost imperceptible nuance.

Being creative is to put new life into things—in our daily lives as much as in art, music, and literature. When we feel depressed, dull, fed up, and worthless, our creativity is blocked, absent, or not being expressed in every breath we take. Sometimes, like the painter Gauguin, we feel we have to break out, leave our home and friends, and voyage to some new land in order to feel totally alive and fresh. Creativity can therefore involve both destruction and rebuilding, and, as mentioned before, the creative act is sometimes blind and brutal. But most of the time, things do not work that way. It is equally possible to remain at the same job, in the same house, and with the same friends but to exist in such a way that, through a tiny shift in our way of encountering the world, everything appears fresh, new, and exciting. After all, much of what we see, hear, and otherwise sense around us comes from within. Our sense of life and the world we inhabit is intentional. It is put together in a creative way so that all that is needed for things to fall in place in a totally novel way may be a minuscule change in perception.

When our perceptions change in this fashion, we are, in effect, transforming the world around us. We are remaking it—making it new. One of the outcomes of this process is that we are also remaking the people around us. When we respond to people in open, creative, and understanding ways, their attitudes toward us change, and the potential is present for something within them to change and open up, in turn.

Old Wine or New Bottles: Renewal or Making It New?

The distinction between renewal and novelty is only relative. At times, artists decide to work within the strict confines of a traditional form as a way of expanding their creativity; at other times, they may feel the need to burst out of the confines of tradition. After all, did Pasolini and Spencer reanimate or create anew particular poetic forms?

The force of creative renewal extends throughout the cosmos. It is present in living systems and colonies of cells. It occurs in the way systems that are close to chaos spontaneously organize and maintain their features. The flow of renewal is an essential feature of a tree, the vortex in a river, and even a modern city.

The desire to renew the world lies at the basis of many traditional ceremonies, such as those carried out daily and seasonally within Native America. Across the North American continent, there is always someone at prayer with a pipe and sacred tobacco when the sun rises. Among the eastern woodlands tribes, there are ceremonies for the first thunder and for the planting of corn, squash, and beans. Plains people hold the Sun Dance each year. Every ceremony creates anew and celebrates the relationship that Native Americans have with the spirits and energies of the land.

People in the industrialized world have almost forgotten the importance of such ceremonies. Yet the shadows of these celebrations can be found in New Year's, Passover, Easter, Octoberfest, Thanksgiving, Hanukkah, and Christmas activities, as well as the New Year festivities of Chinese communities, Burns Night for the Scottish, and so on. People may not think about the seasonal origins of these festivals, but deep down, they understand that, for some reason, it is important to make an effort so that family and friends can be together to mark these passages in the year in some special way—be it having a ceremonial meal together, exchanging gifts, getting dressed in one's best clothes, or simply singing together.

Indigenous people know that these ceremonies are a vital way of keeping a society whole. All around us in our modern world, things decay, rust, erode, and fall to pieces. The only things that manage to survive are living organisms whose inner structures are constantly being renewed. Native people, viewing the world around them, reason that human society is no different: It too is maintained, generation after generation, by acts of renewal. When those traditions—those ceremonies and markers in time—have been eroded, then the society itself begins to disintegrate. And so, renewal begins to merge into healing—for society, the individual, and the world.

In the latter part of the nineteenth century, Gerard Manley Hopkins sought to return poetry to its origins in English speech. To achieve this, he developed a new poetic tool that he called "sprung rhythm." To the eye, this looked like very new poetry indeed, yet to the ear, it sounded more natural than the formal poetry that had gone before.

In the early decades of the twentieth century, Picasso was the icon of the new, for his artistic energy was bursting out of existing bounds. Yet, at the same time, he paid homage to his great predecessor, Cézanne. The series of paintings and sculptures he made toward the end of his life renew the work not only of Cézanne but also of Giorgione and Velàsquez.

In music, Wagner and Debussy felt limited by the requirements of a tonal center in music and began to experiment with alternatives. After hearing oriental music, Debussy began to use a whole tone scale in his compositions. This form of new musical freedom allowed him to write impressionistic pieces that floated free from a tonal foundation. In turn, Schoenberg codified the experiments of Debussy and Wagner into serialism and its twelve-note row and atonality. Now music was constructed out of a basic twelve notes (sometimes, a lesser number, such as a pentatonic row, was used) played in a particular sequence. These notes could be sounded always in the same order but in reverse, upside down, transposed to higher or lower registers, and in any combination of these transformations. The result was a totally new but nevertheless strict grammar for music. It created music without any sense of a tonal center or key, and in its extreme form, even the duration of notes and the instruments on which they were played, as well as their loudness or softness, were determined by such musical formulas.

For a time, composers were keen to experiment with the new form, but many eventually began to feel that they had given up one straitjacket—tonality—in favor of another—atonality. Some of the things they wanted to do in music were more difficult to achieve through strict atonality. Serial music allows composers to exploit complex internal musical orders, but it lacks an immediate melodic quality. (After a concert of atonal music, the audience is unlikely to come out whistling.) Although its internal shapes may be intellectually satisfying, they are more difficult to detect by the ear, so there is always the danger that a composer's ingenuity will go unrecognized. And so, a new generation of composers began to experiment with a return to tonality and to investigate pure sound, electronic production, and fusions of composed, jazz, and popular music, and they wrote in the hypnotically popular style of minimalism.

Steve Reich was one of the developers of this minimalist school of music. In the 1960s, minimalism represented a new form in which he and other composers could explore their particular creativity. Then, in the mid-1990s, Reich returned for his inspiration to the twelfth-century practices of Perotin and the school of Paris. Musical minimalism still remained a creative lode to be mined, but the particular things Reich wanted to do in this period demanded different forms.

The tensions between "making it new" and renewal can also arise when people interact with the material structure of the world; thus, the limitless possibilities inherent in creativity must intersect with the constraints of matter. Let us take the example of the explosion of creativity and invention that took place in the small English village of Coalbrookdale in the early eighteenth century.

Since Neolithic times, high-quality iron had been made by melting iron ore inside a charcoal furnace. At the high temperatures generated in the center of a furnace, the oxygen in the oxides of iron combined with carbon from the charcoal to form carbon monoxide and carbon dioxide. The iron freed from its oxygen then flowed to the bottom of the furnace, where it collected as a lump of pure metal. This was the tradition by which iron was made all over the world.

The disadvantage of the method was that iron could only be made in small quantities. In addition, producing charcoal by burning wood was a slow process, and, as the size of England's forests diminished, the price of charcoal rose. Iron could also be produced in coal-burning furnaces, but there, the impurities in the coal (mainly sulfur) combined with the iron to produce a brittle metal of strictly limited use.

Then, in 1709, Abraham Darby discovered a new way to make good-quality iron in large quantities. While experimenting at Coalbrookdale, near the Severn River, Darby discovered that impurities in coal could be eliminated by first turning the coal into coke and then adding limestone to the coke–iron ore mix. In this way, Darby, his son, and his grandson (both also called Abraham) set the cornerstone for the Industrial Revolution. Just four years after Darby had developed his new method of iron smelting, Thomas Newcomen invented the first steam engine. The iron boilers for Newcomen's engines were cast at Coalbrookdale, and a new source of power for driving machines in factories and pumping water from mines was suddenly available. Then, seventy years after Darby's original discovery and Newcomen's invention, Abraham Darby III built the first railway locomotive at Coalbrookdale for the inventor Richard Trevithick, again using iron.

Darby's iron was available in quantity, and it was used, among other things, in large cooking vessels known as "missionary pots" that were exported all over the world (like those shown in early cartoons of "cannibals" gathered around a cauldron). The Darby family also created the world's first industrial village, paying attention to every aspect of life, providing clean and comfortable housing, properly metaled roads, a pleasant tree-lined walk to work, and an evening college where workers could continue their educations. All aspects of life were planned—except for one. The Darbys simply did not foresee the implications of their inventive creativity—an Industrial Revolution that would sweep across England and totally transform society in the north and the Midlands, turning villages into great manufacturing towns such as Birmingham and Manchester and displacing people from their traditional lives. How different the terrible slums and disgraceful working conditions of the nineteenth century were from Darby's original, philanthropic dream. Again, creativity by itself can be a blind force, and it therefore needs to be tempered by an intelligent awareness of the whole of a situation. (Things did not move so quickly in the United States. The forests appeared inexhaustible, and so charcoal was sufficiently abundant to obviate the need to rapidly adopt Darby's coke furnaces.)

Darby's invention began with the dream of bettering society. It ended with the evils of the Industrial Revolution. So often, creativity involves that Faustian bargain mentioned earlier. It may indeed be possible to sup with the devil; if we use a long enough spoon and insist on a cast-iron contract, we may escape with our souls and ethics intact. But all too frequently, the seduction of money, power, and advancement causes us to make a series of small compromises that, taken together, end up as a very large sellout.

Leni Riefenstahl was an exceptional filmmaker with a wonderful eye. But ultimately, the chance to practice her art to its fullest extent took precedence over her morals and ethics, and so she used her talents for the Third Reich. In terms of visual effects, she produced some marvelous films for the Nazis, such as *The Triumph of the Will* and *Olympische Speil 1936* (which another director, Ken Russell, included in his top-twenty list of films, along with works by Orson Welles, Truffaut, Cocteau, and Eisenstein). Ironically, Riefenstahl would earn the disapproval of Hitler, for in the latter film, she allowed her camera to spend too much time caressing the body of the black American athlete Jesse Owens.

Riefenstahl's work had been irreversibly tainted. If she had not struck such a bargain with the Third Reich, she would have been able to continue working after the war and would probably have been recognized as one of the giants of cinema. (Ironically, a somewhat similar fate awaited another great talent who, unlike Riefenstahl, retained his integrity—Orson Welles. Having attacked the image of William Randolph Hearst in his film *Citizen Kane*, he was blocked from active professional life as a Hollywood director.)

Throughout history, artists have put their talents at the service of corrupt regimes in order to create. They have taken money from patrons and organizations whose ethics and actions do not bear examination. And often, they have felt that their decisions can be justified on the grounds that an artist is above common morality and that their first allegiance is to their art. In a certain sense, there may be some truth to this claim of the Nietzschean *über mensch*. Yet, like the electron that is both wave and particle, the artist is also a human being, a citizen who lives in a corrupt world and must stand up and be counted. Sometimes, as with the composer Shostakovich, an artist must walk a fine line in order to balance art and common humanity.

Such dilemmas are not confined to the arts. There is evidence that medical science has made use of the results of experiments performed on live and unanesthetized prisoners by the Germans and Japanese during World War II.

A particular irony is provided by the Manhattan Project that built the atomic bomb. The political impetus for that project was based on the reasonable assumption that Germany would be building its own nuclear weapon. But even as the tide of war turned and the fall of Germany became imminent, work on the bomb continued. By this time, the atmosphere within the project had become all-consuming for the participants; scientists were working together, cooperating on solving significant problems, and showing how important they could be to a nation's security. As Richard Feynman observed after the war, the excitement of working on such a scientific problem became the major driving force, and many scientists simply ignored (or forgot about) the implications of their work.

The meaning of what they were doing only hit them when they saw the first photographs of the bomb's effects on Japanese civilians. At this point, some of the scientists began to feel guilty—not so much because of what they had done but because they had not previously felt guilt for what they were doing! Oppenheimer remarked that science now knew "original sin."

Scientists, doctors, and engineers have always used their creative skills in time of war to design new and better ways of delivering death and destruction—and not only in conventional wartime. Much of the scientific work in North America during the cold war decades was funded directly and indirectly by the military. Advances in computing power, artificial intelligence, communications satellites, space travel, and even an understanding of bird migration (as a potential vehicle for germ warfare) all resulted from a Faustian pact that involved, on the one hand, the exercise of scientific creativity and the love of knowledge for its own sake and, on the other, the notion that military power must constantly be enhanced. As Karl von Clausewitz put it, "War is nothing but the continuation of politics with the admixture of other means."

Creative Perception

Creativity is ubiquitous. We can't get away from it. But as we saw in the previous section, it can be seductive. To use it wisely and effectively, we need to exercise creativity in our perception of a situation and in our own awareness of how we see the world.

This pertains both to the ethical use of creativity and to the way in which the work we do fits into the actuality of the world. A practical example, again from the Darby family, provides an illustration. As mentioned, their ironwork was located close to the Severn River, which had to be bridged. Previously, bridges were made of stone or wood, but with the availability of iron, people asked, Why not construct an iron bridge? The bridge itself was designed by Thomas Prichard and had an arch 140 feet in length. In 1777, when it was completed, Iron Bridge was seen as a triumph of the new age of iron; today, it is a British national monument. Also known as Coalbrookdale Bridge, it is the ancestor of all the modern steel bridges that span ever greater distances. However, in one significant respect, it differed from everything that came after it.

Up to that point, the designers and craftspeople who had built bridges using stone or wood understood the strengths and weaknesses of these materials, and, ironically, they brought some of that knowledge to the building of Iron Bridge. To our modern eyes, the structure looks like a hybrid: In design, it is a wooden bridge, but it is built with iron—right down to the carpenter's joints used to connect pieces of metal together.

Although Iron Bridge was a triumph that looked to the future, it also looked back to the past, for those who designed and built it had not yet learned to exploit the full potential of the material. Nature had provided them with a new resource, but their consciousness was still working in old ways.

The Iron Bridge example gives us another insight into creativity—the nature of creative perception. Those who first worked in cast iron used skills developed with other materials. It required an act of creative perception to see afresh what was significant in the old approach and what was inappropriate, what should be preserved and what should be exploited in new ways.

Something quite similar occurred with the invention of the internal combustion engine. The first automobiles were appropriately called "horseless carriages," for the pulling power of the horse was replaced in these vehicles by the motive power of a gasoline engine. In the early years of the automobile, designs perfected for horse-drawn power were adapted to serve a new means of propulsion. What was needed were inventors who could turn their imaginations to all the potential of a new material (such as cast iron) or a new source of power (such as the internal combustion engine) and then realize their ideas in the form of new designs.

Take, for example, Neils Bohr's theory of the atom, developed in the early years of the twentieth century. Bohr tried to hold the old and the new together as he devised a theory to explain the stability of atoms. He had studied at Manchester University with the experimental physicist Ernest Rutherford, who discovered that the atom consists of a cloud of electrons around a central nucleus. This led him to picture the atom as a solar system in miniature, with negatively charged electrons orbiting around a positively charged nucleus. The problem with this model is that, according to classical physics, the electron (because of its electrical charge) should be radiating away energy as it rotates around the nucleus. In other words, the electron should lose energy and spiral inward to the nucleus. Atoms should not be stable; they should have only extremely short lives.

The central question Bohr posed was this: Why is matter stable? The answer, he felt, must come from Planck's discovery of the quantum. Planck and Einstein had shown that energy comes in finite packets. Maybe, Bohr reasoned, energy simply cannot leak away from atoms. Unlike planets moving around a sun, electrons cannot choose any orbit they like but are only allowed to move in certain stable—or quantized—orbits. At one stroke, Bohr's theory explained both the stability of matter and the way atoms arrange themselves according to the periodic table of the elements (more about this in Chapter 3). What's more, his theory could be used to calculate the spectrum of light emitted by excited atoms.

However, a young student at Munich, Wolfgang Pauli, was unhappy with Bohr's approach. While others looked at its triumphs, Pauli viewed it in much the same way as we now regard Iron Bridge. The theory grafted new ideas about quanta onto an older structure of planetary orbits. Pauli warned his friend Werner Heisenberg that Bohr had simply put new wine into old bottles.

Pauli's challenge—to discover a new bottle—irked Heisenberg to the point that he fell ill with a serious case of hay fever. Only when he escaped from the university at Munich and took a vacation did a new insight come to him. His totally different approach owed nothing to the past; today, it is known as quantum mechanics. It took an act of creative perception on the part of Pauli to see what was inappropriate in Bohr's model and an act of creative invention on the part of Heisenberg to discover the new theory.

One curious combination of the old and the new is the keyboard on which I'm entering this book into my PC. When typewriters were first developed, they were relatively clumsy mechanical devices with a tendency to jam if the keys were pressed too rapidly in succession. The solution was to redesign the keyboard, placing letters of the alphabet in a way that made fast typing as difficult as possible. In this manner, the speed of typing was slowed and the machines did not jam. As a result, several generations of typists have been forced to learn an unnatural way of typing that was originally designed to slow them down even as they practiced to improve their speed!

With today's PCs, there is no limit to a typist's speed, but an outmoded arrangement of letters (e.g., Q-W-E-R-T-Y-U-I-O-P) is nonetheless found on the keyboards attached to the latest and fastest computers. Although the keyboard is a fossil that should long ago have been replaced, using it is a habit that has become physically ingrained in successive generations of people. (Incidentally, there is also no need for a key to make a little click as it is depressed. The click is included because the first computer users, who graduated from typewriters, felt disoriented without the accompanying sound.) One day, an enterprising manufacturer will release a more efficient keyboard, and the rest of us will take a few weeks to break old habits and learn typing anew.

Much of what we do and the way we see the world is conditioned in this way. We develop skills and early perceptions that turn into habits. It truly requires an act of creative perception to dissolve the old forms of mental conditioning and make way for the new—the first of the criteria for creativity that we examined earlier.

Along similar lines, the story of a gardener who successfully exploited a new material is illustrative. Employed by the Duke of Devonshire, Joseph Paxton turned his hand to building a conservatory and a lily house for the duke as a sideline. His inspiration apparently came not from conventional engineering or architecture using wood and stone but from the natural forms of the plants and trees that grew all around him. Because he was a gardener, he probably was much freer from the assumptions and habits of mind of more orthodox architects and engineers.

In the mid–nineteenth century, organizers of London's Great Exhibition needed a large hall for the many exhibits that would be included, and they finally chose one of the rather cumbersome models submitted by professional architects. Only later did the committee became aware of what Paxton was proposing—a vast, prefabricated building consisting of glass walls with a slender skeleton of iron, something like a great structure of palms. Thanks to the perception of Isimbard Kingdom Brunel and Robert Stephenson, the original decision was overruled and Paxton's innovative idea was accepted. The result—the famous Crystal Palace—was a structure of glass and thin iron columns that covered an area the size of Saint

Peter's Cathedral in Rome. (The final design calculations were made in collaboration with the building's contractors, and, in addition to the idea of a prefabricated building, Paxton introduced the innovation of joining iron members by using hot rivets.)

Another innovation and an act of creative perception in terms of exploiting possibilities followed Bessemer and Siemens-Martin's development of large-scale steelmaking. This time, an engineer was ready to exploit the new material, and in a surprisingly short time, Eiffel built the great tower that still dominates the Parisian skyline.

The creativity inherent in works like these requires an ability to suspend old forms and habits and see things as they really are. This also involves the ability to see things as an organic whole, rather than as a set of fragments. Such was the particular ability of Isambard Kingdom Brunel, an inventor of genius whose vision was so farsighted that well over a century later, most cities and countries still haven't caught up with its potential.

In addition to recognizing the innovative nature of Joseph Paxton's design for the Crystal Palace, Brunel possessed skills ranging from engineering to civil planning. He designed the first true transatlantic steamer (the *Great Western*), the Menai suspension bridge over the Avon River, a tunnel under the Thames, and a prefabricated hospital, and he introduced innovative construction methods such as the use of compressed-air caissons to sink the underwater foundations for bridges. He also built railways all over the world.

Each one of his many inventions would have earned him a place in history. But Brunel's vision went even further: If he could build iron ships, railways, and bridges, why not a totally integrated transport system? Brunel's great plan began at Paddington Station in London. The system he designed permitted Londoners to drive into the station in their carriages and transfer themselves and their luggage directly to the train. Brunel's railway then took them across the Avon River on his magnificent suspension bridge and on to Temple Meads Station, which he built at Bristol. From there, they could board his newest ship, the *Great Eastern*. Powered by screws and paddles and boasting a double iron hull, the ship was unsurpassed for the next forty years. The travelers' next stop was New York.

Brunel's transport system is an example of the integrated vision needed for great, creative projects. It would have been too easy for him to see his various projects as a series of separate inventions; his gift was in understanding how they all fitted together. Put another way, his vision was holistic—but not like that nebulous, flaky holism in which everything is gathered into a featureless soup. A true holistic vision appreciates the organic way in which things fit together or flow one out of the other, while preserving their unique individuality, their particular presence, and the ability of the whole to differentiate in complex and interesting ways.

It is ironic that some 130 years later, few governments and civic planners have been able to equal Brunel's vision of an integrated transport system in which passengers can step from one form of transport onto another without enduring long, unnecessary waits and journeys on foot. (Consider how rapidly you can fly between far-flung cities in the United States and how long it takes to leave the city center, get to the airport, check in, and wait for takeoff. In some cases, the trip to the airport may be longer than the flight itself.) We can only regret that Brunel is not alive today, for he had both the genius of an inventor and the personal authority needed to cut though red tape and get a complicated and innovative project completed.

Brunel's example provides another lesson in creativity: Invention or the discovery of novelty by itself is not enough but must be linked to an overall vision of how that invention or discovery locks into a much wider social or environmental picture. It is often said that new ideas are "ten a penny." What is really important is the ability to follow through, to get ideas realized, and to deal with all the obstacles that are put in the way. In the end, visionaries must be extremely practical people; they have to get their hands dirty and understand the way the world works. This statement applies as much to a great feat of engineering as to a poem, painting, novel, piece of music, or human relationship.

Creation and Society

Whether an inventor, artist, or writer must exercise creativity from within the limitations of a strict form or whether he or she can burst out to create something radically different may not always be a matter of personal choice. Sometimes, societies are particularly stable (as was the case in classical China), and during such periods, creative forms are largely unchanged from generation to generation. There are also times when it seems that the entire consciousness of an age changes and that society demands something dramatically different, in keeping with its new perceptions.

Such a transformation took place in Europe during the nineteenth century, a time of revolution. In France and America, the old political regimes had been overthrown; in Germany and Britain, change was focused on art, literature, and philosophy, as well as engineering and technology. The overthrow of classicism demanded fresh creative forms, and the new age of romanticism was personified in Beethoven's heroic paeans to the human spirit. In literature, this creative spring sustained Goethe, Byron, and Shelley. In art, it nurtured Caspar David Friedrich, Philipp Otto Runge, and Eugène Delacroix.

The classical forms had resulted in great art and literature. At the same time, they were associated with conservatism and the status quo in political philosophy. As the new consciousness of Europe embraced the concepts of liberty, equality, and fraternity, the old had to be swept away and a new consciousness devoted to human freedom had to be adopted. Yet, in embracing the impulses of the unconscious and rejecting the strict forms of classicism, romanticism was left unprotected against the darker and more destructive side of human nature, for if creativity lies in dreams, then dreaming can be enhanced by laudanum and alcohol. Many of the romantics died young, and romanticism itself paved the way for something more excessive than its originators had imagined.

The successors to romanticism, with its emphasis on instinct and the power of the unconscious, ranged from dada and surrealism to futurism and fascism. At the start of the twentieth century, the Italian artist Marinetti issued his futurist manifesto praising the new machines—the airplane, automobile, and, more ominously, machine gun. It extolled speed, force, and violence and even went as far as eulogizing the glories of war. Futurism in itself was highly creative. It introduced speed and dynamism into art. It played with language using free forms, liberating words from syntax, inventing the first artistic happenings, and bringing street and factory noises into the concert hall and thereby giving all sound the potential to be music. But how easily did the rhetoric of the manifestos of futurism slide into the language of fascism. Extreme nationalism, praise of the power of the gun and violent overthrow, and the intoxication of energy came as easily from the mouth of a Mussolini as from a Marinetti.

Both romanticism and classicism have their dark sides, the shadows cast by their light. Once again, we see how creativity by itself can be a blind and amoral force. It is an energy that pushes us forward, like a high wind in a storm, and it must be tempered by another form of creativity—a clear perception of the situation as a whole, of where things are moving and how they will evolve. This anticipates our third form of creativity— that of healing and making things whole.

The movement of consciousness between the intoxication of new freedoms, the creative challenge of traditional forms, and the constraints of the physical world brings us to another theme that will thread its way through this book—the encounter with and impact of the Greek gods Apollo and Dionysus, the two living forces in all creativity. In *The Birth of Tragedy*, the philosopher Friedrich Nietzsche argued that these gods underlie Western civilization: Apollo representing concern with clarity, order, logic, and purity of form, and Dionysus representing intoxication, the power of the senses, and the forces of the unconscious. Yet, as contemporary physics shows, order and chaos or law and chance turn out to be not so much diametrically opposed as partners within a cosmic dance, a dance in which one keeps changing into the other. Thus, Dionysian chaos underlies Apollonian order and simplicity, and chaos is born out of Apollonian forms and symmetries. We shall meet the gods again in the following chapter.

The Postmodern Condition

As I write this book, postmodernism still remains relatively fashionable, with artists free to adapt any style to their own particular uses. A given work may be an amalgam of patterns, forms, and templates. (This is easily seen in postmodern architecture, in which various architectural forms of the past, such as classic columns, exist side by side with glass and steel.) Postmodern novels can be written in a wide variety of styles; their common ancestor is Joyce's *Ulysses,* in which successive chapters are written in the style of newspaper articles, the Bible, romantic novels, and Middle English.

At its best, postmodernism enables the writer to engage the reader at several levels, ironically using one level to address the reader directly and another to deconstruct that form of address and even the acts of writing and reading themselves. Italo Calvino's *If on a Winter's Night a Traveler . . .* is, in part, the author's comment to the reader about the book he or she is presently reading. It is also actual chapters in—what? A book called *If on a Winter's Night . . .* ? Or is it possibly made up of different chapters from some other book? Or quite different books that have slipped within the same covers?

For many writers, musicians, and architects, postmodernism represents a creative form in accord with our present sensibilities about relativism. At the same time, it can be a barrier to a more direct encounter and a symptom of our modern fear of commitment. We have been made painfully aware of the colonialism, fascism, sexism, and racism of our forebears. We realize that our ancestors were largely unaware of the prejudices they held. Today, when any path is judged to be equally deserving as any other, how can we throw ourselves in only one direction with total energy? To do so, we may temporarily have to blind ourselves to the alternatives. And so, we prefer to keep our options open and both Apollo and Dionysus at a distance.

We lack that naive confidence in our own powers that was possessed by the English Victorians (with their sweeping engineering triumphs), the Florentines of the Renaissance, and the modernists of the early part of the twentieth century. We are determined to do no wrong, but without the breadth of energy that is simultaneously iconoclastic and respectful of the triumphs of the past, we are unable to produce great works of art that will inspire our children's children. The truly heroic does not accord well with our age. And the result can be a form of paralysis and a drying up of our creative springs. All periods need their artists, and all societies require creative acts of renewal. Consequently, if we feel our own age is somehow lacking and our shared sense of meaning is not as enriching as it should be, this is, in a certain sense, the sickness of our age. This is not to say that our times are in need of some millennium-spanning work of art, something on the grand and heroic scale. Creativity is badly needed, but its products no longer have to be manifest in song, stone, or verse. Maybe what is required are works of art that exist at the social level— an opera of meaning, a poem of human interaction, a painting of belief.

Creativity as Healing

This brings us to the final way of thinking about creativity, which is as an action of healing. At first sight, this may not be a familiar way of thinking about creativity, but it turns out to be one of the most important.

One of the functions of healing is to restore and make whole. This wholeness involves our minds and bodies, our family and friends, the society in which we live, and, ultimately, our planet Earth and our connections to the entire cosmos and the transcendent. But how do we know when healing is necessary? How do we know when something is fragmented? How do we know when it has finally been made whole? To grasp these things requires a highly intelligent act of perception and discrimination, and, as I shall emphasize in this book, all perception is itself an essentially creative process.

It's pretty obvious that when a system has been broken into disconnected fragments, it is necessary to bring it together again and heal its internal divisions. But healing is also required when what appears on the surface to be whole is, in fact, made of elements put together in arbitrary and unthinking ways.

The colonial empires of the past drew arbitrary boundaries on their maps to define new countries and protectorates. This was done without any consultation with or deep understanding of the peoples who had inhabited those regions, the way they lived, their histories, and their particular connections to the land. The result was the creation of artificial nations, held together by a colonial bureaucracy confronting great internal tensions. In many cases, after the colonial government withdrew, such countries were torn apart by conflict. The attempt to create a whole had been done in an arbitrary fashion, without any attempt to involve the histories and values of the affected populations. The result was an artificial whole racked with tensions and contradictions. Just as making whole requires the removal of arbitrary divisions and the forging of new links, sometimes an act of healing means that things should be taken apart and given their own autonomy.

Such notions of healing and wholeness extend to our bodies. Living systems have chosen to stand poised on a knife-edge. A living system must be open to its environment. If it is isolated in a box, it will soon die. Life requires a constant exchange of energy and matter with the outside world. Yet this same openness renders a system vulnerable to changes and invasions from outside.

Something analogous happens with a computer. Used by itself at home, it will continue to perform its tasks perfectly— one hopes. But to make it even more useful, it must be connected to the Internet and augmented by downloaded files and software. Once a computer is connected to the outside world, however, there is always a danger that it will pick up a virus that will wipe out all the information stored on the hard drive. The solution is to protect the computer with an antivirus program.

A computer and the viruses that attack it are relatively simple when compared to the human body and the range of viruses, microorganisms, poisons, and radiation that surround it. This is why we have an immune system, a system whose complexity is comparable to the human brain. To work effectively, the immune system has to be intelligent. It has to know what it means for the body to be whole and healthy. It must discriminate between what is needed from the environment and what it must attack. Most important, it must refrain from attacking the essential material of the body it protects by recognizing friend from foe.

This metaphor of healing can be extended to the body of society as a whole, embracing the divisions its members experience between body and mind. It is present in the merging of our personal horizons with the transcendent. Perhaps it is for this reason that the arts are so highly valued, for art—from Gregorian chants to Easter Island statues, from Fra Angelico to Charlie Parker, from African masks to John Donne, from Blackfoot traveling songs to Oedipus Rex—puts us in touch with the transcendent. For a time at least, art frees us from our restrictive preoccupation with the secular and connects us to the eternal. In its healing function, art can be new and innovative, it can renew and revive, and it can connect and heal.

And so, creativity, as expressed in art, nature, society, and the lives of each one of us, is a constant play of three aspects. The force of creativity connects us to the transcendent and unlimited and thereby heals. It acts to renew and revitalize our selves, our society, and the world at large. And finally, it brings about the new and the unexpected and fills our lives with eternal surprise. In the next chapter, we will learn a little more about this creative force that pervades the natural world, as well as our own relationship to it.

chapter 2

The Living Universe

"The force that through the green fuse drives the flower."
—Dylan Thomas, *Collected Poems 1934–1952 (1952)*

It is too easy to think of creativity as something only associated with poets, artists, composers, filmmakers, designers, outstanding scientists, and those rare individuals we call geniuses. This book argues that creativity is possessed by all of us. It is deeply embedded within our bodies and extends throughout the material world, from the atom to the big bang theory of the origin of all that is.

Think of the quotation that opens this chapter: "the force that through the green fuse drives the flower." This line, from one of Dylan Thomas's earliest poems, reveals the poet's deep and mystical sense of connection to nature. This theme runs through much of his poetry until his final works, such as "In the white giant's thigh" and "Poem on his birthday."

We are familiar with the image of Thomas as the drunken stage-Welshman, wenching his way across America to meet his death by booze in New York City. Less well known is Thomas the craftsman, who only wrote when he was stone-cold sober and in the little wooden shed above his house in the village of Larne, in South Wales. He called this shed his "poem room" and remarked that its window looked over the "heron priested shore" where "cormorants scud" above the "congered waves" of Carmarthan Bay (from "Poem on his birthday").

Like any poet, Dylan Thomas was a wordman who spent days in his poem room struggling to find just the right word for a single line of verse. His friend Vernon Watkins describes the way Thomas used separate work sheets to "build the poem up, phrase by phrase, at glacier-like speed" (Foreword to Dylan Thomas, *Adventures in the Skin Trade*, 1955). In the best of Thomas's work, each word reinforces and complements every other. Take the single line above and the words *force, fuse*, and *drives*, which suggest dynamism and an unrestrained potency but one that must be channeled though a *green fuse*—an image that also suggests the need for compression into a narrow channel (the birth canal, perhaps) before it can burst into fruition.

Thomas's nature poems have no gamboling lambs or pretty sunsets. They are about the naked power of nature, that power inherent in every living thing, from the flower to the poet himself. It is a power that lies beyond good and evil, for Thomas knew that, in nature, "finches fly in the claw tracks of hawks" and the "rippled seals streak down to kill" (from "Poem on his birthday"). Nature is alive with this profligate power. It compels the eagle to hunt, the root to break rocks, and fungi to force their way through the paving stones on a sidewalk.

In this chapter, we shall explore the creativity of the natural world and the way we have distanced ourselves from that source. Later in this book, we shall look at our own relationship to nature and the world that lives within our bodies. But first, we must understand the way a certain attitude that began in the thirteenth century has caused us to distance ourselves from the natural world.

Before we begin, one warning is in order. Nature absorbs all our definitions and analyses within itself. It transcends anything that we attempt to define, that poets try to scan, or that artists strive to limn, and still it forgives us. In the end, there is no dark or light side to nature. These are categories we create in our attempts to understand the world around us. Nature is simply there, ever present, waiting. As Camille Paglia says in the opening sentences *of Sexual Personae: Art and Decadence from Neferitti to Emily Dickinson* (1992), "In the beginning was nature ... we are merely one of a multitude of species upon which nature indiscriminately exerts its forces. Nature has a master agenda we can only dimly know."

Dylan Thomas came from a long tradition of English (or, more precisely, British) writing that celebrates our connectedness to nature. That tradition is present in the first poems written in what is recognizably the English language. *Sir Gawain and the Green Knight*, from the fourteenth century, contains vivid descriptions of a winter travel to the dark north of England, as well as boar-hunting scenes at Christmas. From the same period comes the dream landscape in *Piers Plowman*.

Although very much an urban writer, Shakespeare was also aware of the effect of landscape on the human mind. Nature is celebrated in the writing of Andrew Marvell, Wordsworth, and the Lake Poets Gerard Manley Hopkins, Thomas Hardy, and D. H. Lawrence, as well as in the works of more contemporary writers, such as Ted Hughes, Thom Gunn, and Nobel Prize winner Seamus Heaney.

Maybe this tradition is the result of Britain's compact size and the fact that, until the Industrial Revolution at least, small rural societies had existed unchanged for centuries. The country has a gentle landscape, mostly farmland, without extremes of climate and terrain. People of this land are keenly aware of their heritage; indeed, quite ordinary and normally unpolitical individuals will turn out on a wet Sunday morning to protest the cutting down of an ancient oak or the closing of a footpath.

England's landscape encourages intense and close observation of detail, and its writers often display an almost mystical sense of connectedness with the land. This intimacy with nature also extends to the work of English painters, from the Elizabethan miniaturists through Thomas Gainsborough, John Constable, W.J.M. Turner, and Richard Wilson and on to contemporary land artists such as Richard Long, Hamish Fulton, and Andy Goldsworthy.

This sense of the immanence of nature is present in the work of the eighteenth-century painter Samuel Palmer. His early paintings are all on a small scale, yet they are curiously intense.

The poet and engraver William Blake had criticized the New-tonian vision of the world for overpriviliging reason and analy-sis and omitting the immediacy and transcendence of our experiences. Inspired by Blake, Palmer moved to Shoreham in Kent with a group of friends called "The Ancients." Shoreham became their "valley of vision," where they would paint, read together, and talk late into the night, to the point at which the local villagers believed they were magicians or astrologers.

While in Shoreham, Palmer experienced nature with a par-ticular intensity. Writing to a friend, he said, "Great hopes mount high above the shelter of the probable and the proper; [they] know many a disastrous cross wind and cloud; and are sometimes dazzled and overwhelmed as they approach the sun; sometimes, vext and baffled, they beat about under a swooping pall of confounding darkness; and sometimes strug-gle in the meshes, or grope under the doleful wings of tempta-tion or despair; but shall scape again and once sing in the eye of morning" (Palmer to George Richmond, quoted in Ray-mond Lister, *The Paintings of Samuel Palmer*, 1985).

For the next few years, Palmer's artistic and spiritual vision was of a world in which everything was alive. There were times "when inspir'd by art I am quite insensible to cold, hunger and bodily fatigue, and have often been surprised, on turning from work to find the fingers aching and nearly motionless with in-tense cold" (quoted in Lister, *The Paintings of Samuel Palmer*, 1985).

His sense of the transcendental fruitfulness of nature can be seen in a number of his paintings of the period, such as *The Magic Apple Tree* and *Coming from Evening Church*. *A Hilly Scene* depicts his "valley of vision" under moonlight, with cornfield, tree, and church spire. The field is ripe and heavy with autumn corn, yet we notice that the tree, a horse chest-nut, is also laden with white spring blossoms. Now we realize that all the seasons of fruitfulness are occurring simultane-ously. Nature bursts with life, and for Palmer, the source of this life is the light of the divine.

Palmer's mysticism in the face of the creative force of nature was echoed in the twentieth century by Stanley Spenser.

Spenser's paradise was the Thames-side village of Cookham. His painting of apple pickers evokes the fruitfulness of Palmer's *Magic Apple Tree*. As Spenser wrote of his painting, "Places in Cookham seem to be possessed by a sacred presence. . . . The people in 'Apple Gathers' are, as it were, brought forth by the place and therefore, are aware of its divinity. They are expressions of the divinity there presiding" (quoted in Simon Wilson, *British Art from Holbein to the Present Day*, 1979).

Through the intensity of Spenser's inner vision, Cookham is transformed into a paradise, and the New Testament miracles occur in the homey environment of the village, the regatta, and the churchyard. The same natural power that for Dylan Thomas drives though the green fuse of a flower is present in all manner of everyday objects for Spenser. In one of his paintings, a group of women gather around a dustman (a garbage collector), embracing and lifting him in praise of his everyday labor, and the contents of his bin—a cabbage stalk, a broken teapot—are treated as sacred, immanent objects.

This intense contemplation of nature and its creative powers is not confined to the British Isles. In America, one only has to think of Henry Thoreau's meditations at Walden Pond or the close observations of Annie Dillard. From sixteenth-century Spain came Sanchez Cotan's deeply mystical still lifes, and what could be more disarming than a few vegetables hanging by strings against a deep-black background? But these paintings have such a powerful religious intensity that they become an epiphany of what the rest of us generally take for granted. Centuries later and on the other side of the Atlantic, Georgia O'Keeffe was looking with a similar depth at cactuses and flowers, Emily Carr was studying the British Columbia forests, and the Canadian Group of Seven was contemplating Ontario's lakes and trees. Meanwhile, in Holland, Mondrian's identification with the tree took him from naturalism into extreme abstraction.

In all these examples, nature is a living agent, a power, a presence. Moreover, the works have arisen out of the artist's or writer's ability to merge horizons with the natural world. This is the essence of the statement made by Paul Cézanne as he

painted the motif from nature: "The Landscape becomes reflective, human and thinks itself through me. I make it an object, let it project itself and endure within my painting. . . . I
become the subjective consciousness of the landscape, and my
painting becomes its objective consciousness" (quoted in
Richard Shiff, "Cézanne's Physicality: The Politics of Touch,"
in Salim Kemal and Ivan Gaskell, eds., *The Language of Art
History*, 1991).

This sense of the immanence in nature can also be present in
music. The gently rolling geography of the Malvern Hills lies
behind much of what Edward Elgar wrote. A tight sense of observation is present in the "night music" from Béla Bartók's
Concerto for Orchestra, as is a sense of expanse and mystery in
Gustav Mahler's symphonies, of mystery in Olivier Messiaen's
cosmic expanses, and of the underlying menace of a world dis-
turbed in the garden scene of Maurice Ravel's opera *L'Enfant et
les Sortliges*.

In such works, one is opened to a transcendent sense of the
natural world that surrounds us. It is nature that can lure and
intoxicate, nature that can repel and instill fear. In the space of
a day, nature can range from a breathtaking sunrise and the
serenity of a soft, warm morning to a raging thunderstorm or
the terror of a tornado.

Yet, in spite of nature's inherent power, we have increasingly
separated ourselves from the direct, day-to-day experience of
the natural world. We have distanced ourselves to the point
that we only become aware of nature when it roars in anger or
cries in pain. We only seem to know of nature at times of
earthquake, flood, and forest fire or when we are told that
lakes, rivers, forests, and entire species are dying.

Our modern feeling of separation from nature as a living being—which, in the end, means a distancing from the creative
forces of the cosmos—is the result of a long, slow process that
had its origins centuries ago. That process has veiled nature's
inherent creativity from our present, industrial society. But the
sense of nature's inherent power is very much alive within
other societies. The Blackfoot of Montana and Alberta tread
gently on the earth, for they are walking on their mother's

flesh. The "forest people" of Africa sing to the forest to keep it happy and to ensure that it will continue to look after them and not sleep and forget them in its dreams. (The forest people are referred to by Europeans as "pygmies"; see Colin M. Turnbull, *The Forest People*, 1987.) For the ancient Greeks, Earth was Gaia, a living being, and the indigenous people of the Hawaiian islands tell me that they should not dig too deeply into Earth for it is alive.

A similar vision of the natural world was present in Europe until the end of the early Middle Ages. Although miners in that time dug into the earth to extract its minerals, theirs was considered a sacred task, for metals are born in the womb of Earth. Thus, miner, smelter, and smith alike were engaged in the task of assisting nature to come to fruition. Indeed, for centuries and even millennia, human beings all over the world lived in intimate contact with nature. When Native Americans speak of "all my relations," they are referring not only to their blood relatives but also to the plants and animals, insects and water dwellers, rocks and trees, and the sky beings and mother earth. All are deemed equally alive and deserving of respect.

This ancient vision continues among the old people of the hilltop village in Italy where I now live. They recall that in their parents' and grandparents' time, the earth gave them everything. Their medicines grew wild in the woods and beside the paths where they walked. Food could be grown, trapped, or hunted. Clothing and coverings came from sheep on the hillside or the fibers woven from the ginestra plant. Heat and cooking required wood. Oil and wine came from the tree and vine. At every level, nature's inherent creativity was present to aid and teach human beings. Each spring, nature appeared anew. Nature sustained. Nature healed. But this vision shared by artists, poets, and the people of my village is something of an exception in the modern world. The rest of us don't really live that way anymore. For most of us, the world has been objectified and distanced.

This changing view of nature resulted from the accumulation of many different factors that led to an increasingly secular world. By the end of the thirteenth century, Europeans had

begun to view space and time in a very different way than their
ancestors had. Formerly, space and time were united into a
rich matrix. Time was the movement of the seasons, of renewal, and of the cycles of the church's calendars. After the
first clocks began to appear on public buildings, time soon
became quantified and reduced to numbers. In the same period, tradespeople began to keep accurate accounts, and double-entry bookkeeping enabled them to think in terms of
profit and loss. And so, it wasn't long before businesspeople
began to worry about controlling the commercial world
around them, planning for the year to come, and trying to
predict what the future might hold.[1]

Starting in the mid–thirteenth century, Europeans developed
powerful tools of abstraction and ways of manipulating abstractions as objects within the mind. The use of arabic nu-
merals and double-entry bookkeeping, methods of surveying,
ways of representing space on maps, and so on dramatically
increased what people could do from within the comfort of
their own minds. It was no longer necessary for the world to be
immanent or immediately touchable for it to be understood.
Previously, the real was equated with what could be held in the
hand, with what was manifest. But now, what could be com-

---

[1]It is possible to read Shakespeare's *Merchant of Venice* in this light. The
merchants of the Rialto are rational Renaissance men who have learned to
balance profit and loss and to invest their gold in hopes of making financial
gains from a successful trading voyage at sea. The merchant Antonio, confi-
dent that he will be even richer when his ship returns, borrows from the
moneylender Shylock.

Yet nature, both human and elemental, cannot be controlled. Antonio's
ship is lost at sea, and the loan must therefore be paid back. The inherent un-
predictability of an uncontrollable nature is also echoed in the passions of
Shylock, the moneylender who represents all that is exotic, and foreign to
the rational Venetians. If his loan cannot be repaid, then a pound of Antonio's flesh must be sacrificed. Ultimately, Shylock is thwarted and order is restored within the state but only after the anti-Semitism that lies behind
Venetian society's humiliation of the Jew is revealed. In seeking to restore order and harmony, the state exposes the shadow side to reason. That which
has been long repressed in human nature, in favor of the illusions of reason
and control, must also have its day.

puted, seen on a map, read in a book, or contemplated in an engraving became even more real and immediate.

It goes without saying that the peasant in the fields knew none of this. But for the bourgeoisie and the church—all those who lived in the comparative comfort of towns, palaces, and monasteries—the world was abstracted, physically distanced, and smoothed over. Within the city, affairs could be managed and controlled. Outside the city, humans were at the mercy of everything from brigands to wolves, as well as nature's every whim. But even the bear was harmless when brought within the confines of city walls to dance. The city represented civilization, order, and law. Outside nature was increasingly distanced. In Shakespeare's *King Lear*, for instance, the world of the court is contrasted with the storms, wildness, and madness of the heath.[2]

It is often said that science is responsible for the way we have distanced ourselves from nature and come to view it as a controllable machine. But, as we have seen, these tendencies actually originated much earlier. They began with the abstract mental tools first developed in the thirteenth century that allowed educated people to manipulate the world mentally. Such tools enabled people to treat nature as an object that could be moved around in the mind. They allowed them to speculate about alternatives, to plan and control and treat na-

---

[2]The one thing that could not be controlled, apart from human nature, was disease. It is difficult to estimate the effects of the plague (the Black Death) on Europe. A glance at the history of Italian painting reveals a blank space between the end of the 1340s and the start of the 1400s. During that period, a significant percentage of Europe's population disappeared, and people had far more pressing matters to think about than commissioning frescoes. What effect, I wonder, did this scourge have upon European consciousness? Did it perhaps shake the faith in a benevolent nature?

Something similar but on a much vaster scale occurred in America with the coming of the first Europeans as men and animals brought with them diseases for which the people of the New World had no immunity. Some historians felt that the rapid spread of disease produced a crisis in the First Peoples' faith in terms of their long-standing relationships with the energies and spirits of nature. For neither their traditional medicines nor their appeals to the keepers of the animals worked against the new diseases.

ture as something exterior to themselves. By the time the Renaissance dawned, this way of thinking was well established. And it was this mind-set that spawned Western science.

As a way of looking at and thinking about nature, science itself produced a further distancing. But it also gave human beings enormous power. The developments of our modern world would have been impossible without it. Science has stimulated some of the greatest minds to produce a new type of knowledge, as well as theories of considerable aesthetic beauty. Yet, at the same time, it has caused a further cleft between the heart and the head. It is not that science contains some internal defect but that it and its associated technology simply advanced too fast for human beings.

As a methodology, science prides itself on being objective and value-free. Inevitably, it projects these same attributes onto the natural world. Thus, everything from rocks and trees to stars and human sexuality becomes an object for study and analysis. What to the Blackfoot and Ojibwaj of North America is a constant flux is for the scientist a series of events occurring in space and time. In turn, these events can be correlated to reveal underlying orders, patterns, and forms. Because of the regularity of these patterns (which, moreover, exist independent of individual wishes, beliefs, and desires), laws of nature can be posited and future events can be predicted on the basis of our knowledge of the past.

If nature can be analyzed within the mind into a series of connected events, then it immediately comes closer to being a mechanism than an interplay of powers and spirits or an eternal flux. Moreover, if nature is viewed as a mechanism of considerable internal regularity, it is logical to assume that it has been constructed by reason and that such reason cannot be too different from our own. Such a view was acceptable to the church, for it connected to a much earlier strain of theological teaching. Melding the two, the rational mechanism of the cosmos becomes the handiwork of God, the "unmoved mover." Older astronomical observatories often display a text inscribed around their domes exhorting us to survey the heavens and view God's handiwork. In one observatory, for example, words taken from the Book of Common Prayer are used: "The heavens declare the glory of God: and the firmament sheweth his handy-work."

The early church made a sharp distinction between the natural, material world—that which was made according to God's rational plan—and God's own realm. This difference is clearly defined in the Nicene Creed. Because Christian theology has had such a profound effect on the European worldview, it is worth noting the exact definition. Although the natural world was fabricated by God, the Son of God was not made but "begotten" or generated (in Latin, *Genitum non factum*). God's Son and God Himself have a profoundly different ontology from the cosmos. The cosmos had been made in time, but Christ, as the Second Person of the Trinity, lies outside time and has an entirely different nature in being "begotten." Later, God, in the person of Christ, enters the domain of time and history to take on but remain unlimited by a manifest form.

Thus, the essence of the world was considered inferior to our immortal souls. Such a theological position gave new force to the Old Testament dictum that human beings have dominion over the earth. In turn, science was able to reinforce the notion that the world is a mechanism, something made. And science itself, as developed by Galileo, Copernicus, Kepler, Bacon, and Newton, magnified the power of the mind to manipulate the world in thought. It viewed the cosmos as a great mechanism, albeit a subtle one. It suggested that the cosmos had been constructed and that what has once been put together could also be taken apart and analyzed.

Since the rise of science, ordinary humans have tended to live within a somewhat schizophrenic world. They are taught that the cosmos is a constructed mechanism that consists of objects and events obeying fixed laws. If such a world appears to have qualities and values, then it is only because we have projected such concepts onto nature. Yet people also have experienced the immediacy and immanence of the world. In special moments, they are even able to merge horizons with the world and know it to be a living place. And so, their bodies, their intuitions, and their very sense of "being in the world" are telling them one thing while knowledge and reason suggest another.

This schizophrenia even affects scientists themselves. Scientists may well plead objectivity, but this very plea is based on faith and belief. They believe that the universe is ultimately a rational place and that they are not deceiving themselves with their scientific theories. They also have faith that an abyss will not open up at some point to expose an entirely irrational cosmos. They argue that reason, logic, and the language of mathematics are sufficient to comprehend this world. As the physicist David Bohm used to joke, when a hard-nosed scientist boasts "I have no philosophy" or "I don't bother about metaphysics," he or she is, in fact, making a fundamental philosophical and metaphysical assumption that involves a leap of faith in terms of the power of reason and logic.

In no sense are the preceding observations meant to be a criticism of science. Rather, they are offered to make explicit the belief system on which science is based. That system is a methodology that has produced the sorts of intellectual triumphs that poetry and art can never achieve in their respective approaches. Science is one of the glories of the human race and has produced a profound understanding of certain aspects of the natural world. So when we speak of distancing ourselves from nature and experience, it is not so much science that is at fault; rather, the blame lies in the rapidity with which science and technology have outstripped our human abilities to accommodate them.

For this reason, there has always been a tension between the poetic and the scientific visions of nature. William Blake, for example, spoke out against Newton for describing a world that his own mystical vision rejected. When Blake said that he had once seen the soul of a flea, he was not making a scientifically verifiable statement. He was communicating something essential about his relationship with the world. Blake lived in a world replete with meaning and value, a world that extended far beyond what was tangible or measurable in the scientist's laboratory.

The tension between the scientific, rationalistic approach to the natural world and the intuitive response to the forces of nature forms the basis of Herman Melville's short story "The Lightning-Rod Man." In this tale, a lightning-rod salesman arrives at the narrator's cottage at the height of a violent electrical storm. Ostensibly there to sell his wares, the visitor describes in increasing detail the dangers of exposure during a storm and the narrator's perilous situation in his own cottage. The more the narrator makes light of the danger, the more the salesman reveals his own obsessive fears about the natural world—fears that can only be combated through reason and science. To the salesman, the warmth of the hearth spells danger, for lightning is attracted by the hot air of the fire. Even the walls of the house can conduct the current and thereby present a threat. The salesman insists on standing on an insulated rug in the dead center of the room, from which point he exhorts the narrator to guard against the violent forces of nature.

When the narrator exults in the power of the storm, the salesman replies in terms of the superior electrical conductivity of his copper lightning rod and of the number of square feet of protection it will provide. As the tension between them increases, the narrator accuses his opponent of defying the deity; the salesman is taking a false power upon himself if he believes that reason and science alone are able to overcome the powers of nature. At this point, the salesman reveals the mania beneath his mask of rationality. He himself is the lightning king, he announces, attacking the narrator with his lightning rod in an attempt to kill him.

Just as the scientific logic of the salesman gives way to a burst of madness, so too is the narrator's intuitive sense of oneness with the forces of nature transformed into rationalization, and in the weeks that follow, he vainly attempts to warn his neighbors of the dangers inherent in the lightning-rod salesman.

The faces of reason and mania, instinct and science are also the faces of the gods Dionysus and Apollo. As we saw in the previous chapter, Neitzsche had pictured them as the opposing forces that weave their way though Western civilization. They are active today and present in the tensions of our own society, which embraces the mastery of science yet acknowledges the power of the irrational. The physicist Wolfgang Pauli, one of the key figures in the development of quantum theory, argued that physics must ultimately come to terms with what he called "the irrational" inherent in the natural world. Dylan Thomas wrote of the force that drives the flower. At one and the same time, the notion of force is both a cornerstone of physics and a poetic metaphor.

Let us look a little closer at the incarnations of these gods. Apollo is the son of Zeus and, in earlier myths, is also identified as the son of Horus, the sun god of the ancient Egyptians. As well as being the founder of states, a lover of moderation, and the enemy of barbarism, Apollo is associated with the Python prophet. At various times, he has been described as the god of philosophy, mathematics, astronomy, medicine, music, and poetry.

Dionysus, also known as Bacchus, is another son of Zeus and is accompanied by maenads and satyrs. He is the suffering god who died and was brought back to life. (In earlier incarnations, he may have been the sacred king killed and devoured by a goddess.) Reared as a girl and kept in women's quarters until puberty, he is the god of vegetation, fertility, and wine, as well as the inspiration within poetry and music. Dionysus is the god of ecstasy and of the spring festival in which everyone becomes intoxicated with desire. Yet he also has taught the human race the elements of civilization. Impotence is the curse for not respecting his cult.[3]

At Delphi, the two gods reign together. Indeed, they are so entwined in Greek thought as to become inseparable. According to Nietzsche (in *The Birth of Tragedy*), the masterpieces of Greek art are the expression of the Apollonian characteristics of order, balance, and harmony. However, a full understanding of art can only be gained by coming to terms with the Dionysian.

---

[3]Gentle reader, be warned!

These gods represent the difference between a dream and intoxication. In dreams come those forms that find their way into sculpture and poetry. And when we dream, we are truly being artists, for the dream itself is a work of art. By contrast, the terror and ecstasy associated with Dionysian intoxication overwhelms the senses. Although the poet may have seen the gods in a dream, he *becomes* that god with intoxication. He is no longer a human being who produces works of art but has actually become a work of art himself.[4]

The presence of these gods can be found not only in ancient poetry and the arts but also in contemporary physics. (I'll say more about this in the following chapter.) On the one side stand the Apollonian laws of physics and the symmetry principles that govern the transformations of the elementary particles. This order lies beyond the purely temporal and contingent. On the other side is a constant state of becoming in which nothing, not even the laws of nature, can be fixed and static. This state is the Dionysian chaos that occurs when energy floods into a system and makes it random and unpredictable.

The Apollonian world of simplicity and rule appears, at first sight, to be opposed to the Dionysian intoxication of flux and chaos. Yet, as noted earlier, chaos itself is born out of the Apollonian forms and symmetries. Yet, as contemporary physics shows, the basis of Apollonian order and simplicity itself turns out to be Dionysian chaos.

In all mythologies, opposites must, at some deeper level, unite. The British composer Sir Michael Tippet, who had a deep understanding of the powers of the two gods, kept an image of their encounter—a medallion on which their faces are depicted on opposite sides. If creativity is the blind force that drives the flow of nature, then the way it is manifest is revealed in this encounter of Dionysius and Apollo.

The two gods play out their game within Thomas Mann's novella *Death in Venice*. In this work, the protagonist, Gustav

---

[4]The German artist Joseph Beuys looked to the end of art in the sense of its being some separate and distinct function in society. He called instead for a time when "everyone is an artist," a time in which our actions within society become "living sculpture."

von Aschenbach, has achieved fame and honor for his Apollonian writings. His literary style has developed into an almost exaggerated classicism of symmetry, purity, beauty, and simplicity—albeit tinged with a degree of cynicism and detachment about art and life, as well as an intolerance for human weakness.

Despite his fame, however, Aschenbach remains a driven man, one who has lived his life like a closed fist. As the novella opens, he is approaching the last decades of his life. Compelled as ever to work, he realizes that he no longer experiences joy in writing. The solution, he feels, is to take a vacation somewhere in the warm south, away from his native Munich. And so, he ends up approaching Venice from the sea.

As poets and painters have noted, Venice is a city suspended between sea and sky. It is a place in which dualities and distinctions dissolve in the morning mists. Venice is a city of masks, carnivals, and sexual ambiguities—an ideal stage for the gods' encounter. As he disembarks in the Italian city, Aschenbach finds himself slipping into a world whose qualities are markedly different from the classical clarity exemplified by his own writings.

At the Hotel des Bains on the Lido, he notices a handsome young Polish boy named Tadzio. As the days go by, Aschenbach becomes increasingly preoccupied with the youth. At first, he attempts to rationalize his interest by contemplating the pure, uninterested love of an ancient Greek philosopher for a young man, and he dreams of dedicating his latest work to the youth. Only later, when the young man smiles at him, is he willing to admit, "I love you." Now Eros is unleashed on the world, and Aschenbach's initially dispassionate interest becomes an obsession.

Eros and the erotic element of creativity are other threads within this present book. Eros, who later became sentimentalized into Cupid with his arrows of love, may well have been the first of the gods, hatched from the world-egg, or the son of Chaos, the primeval emptiness of the universe. Eros is the fertilizing force that united Heaven and Earth and produced the world out of chaos. As the god of uncontrolled sexual passion, he exerts a disturbing force on ordered society and is sometimes placed in opposition to his brother, Anteros, the god of mutual love. Although later Greek thinkers attempted to water down Eros's passionate nature, he nevertheless survives into our own times as the Freudian libido—the energy that powers us all.

The erotic and physically sensual side of creation is a significant aspect within the lives of a great many artists, writers, and composers. Degas, Picasso, and Toulouse-Lautrec each made a series of brothel etchings, and in Degas's final works, naked bodies are transformed into landscapes. As a mature artist, Cézanne suppressed a savage, almost masochistic vision of women into his paintings; toward the end of his life, the painter avoided being physically touched, even by his friends.

Eros had long been missing from Aschenbach's life and writings. But as he allows the god to enter him, his personality, which had been so one-dimensional, now becomes intoxicated and demonic. While he pursues Tadzio, the very face of Venice seems to change. The sun burns down, the air turns foul and stagnant, and everything is covered with haze. Aschenbach discovers that the plague has entered Venice. He is told that the disease is found deep in savage swamps and jungles—a location far from the Apollonian order of the city and closer to the chaotic side of nature, as so well portrayed in Conrad's mythic *Heart of Darkness*.

Throughout his life to that point, Aschenbach had offered his talent to Apollo, the god of order and harmony. He had been a staunch defender against the forces of chaos and misrule, but now his obsession allows him to trample on reason. At night, he dreams of a voice announcing "the stranger god," and in his sleep, Aschenbach is drawn into an obscene, cannibalistic, yet seductive mystery.

On awaking, he realizes that he has totally surrendered to the stranger god. There can be no more rationalization about a pure, impartial love of age for youth. The writer who had rejected all sympathy with the troubled depths of the human soul now recalls Socrates' words to the young Phaedrus. The poet, Socrates says, is destined to walk side by side with Eros. In the pursuit of beauty, he must exult in passion and wander in the groves of feeling. The poet can never be a model citizen, for creating art remains a dangerous practice. In their pursuit of beauty, form, and detachment, even the noblest minds are eventually led to intoxication and desire.

The novel ends as Aschenbach, who has never uttered a word of greeting to the boy, watches Tadzio walk toward the sea. The writer dies in his chair as the boy steps into the sea and then turns from the waist to look back at him.

Something within Mann's story burst the bounds of the novella. Not only was it made into a film by a major director, Luigi Visconti, it was also the subject of composer Benjamin Britten's last opera, *Death in Venice.* In the opera, Britten makes the stranger god of Aschenbach's dream explicit. Eros has entered the world but now acts through Apollo and Dionysus. The two gods cry out for the poet's allegiance. Will the writer be ruled by reason, form, and law? Or will he join in the dance, entertain the sacrifice, and accept the mysteries along with their "knowledge that forgives"? Aschenbach wakes into total renunciation. He can no longer fight his fate, so he declares, "Let the gods do what they will with me."

Another composer, Michael Tippet, a contemporary of Britten's, told me that this opera, with its encounter of the two gods, is essentially about music itself and the creative act of composition. Every composer is torn between the sensory seduction of sound and the rule of form, balance, and harmony.

Thomas Mann based the character of Aschenbach partly on the composer Gustav Mahler and partly on his own encounter with a Polish boy in the Hotel des Bains at the Lido. But to interpret Mann's story or Britten's opera as no more than an allegory about the acceptance of gay love is to miss a more important point. Both artists deal with themes and powers that are universal.

Even given Mann's interest in the beautiful youth at the Lido, the writer also had in mind Goethe's obsession with a young girl he had met in Marianbad. And though there are homoerotic and sadomasochistic themes in Britten's *Peter Grimes* and *Billy Budd*, the composer's last opera goes beyond these in its theme of seduction by the god Eros and the struggle between the opposing forces of order and sensuality in the mind of an artist.

This subject must have been particularly ironic for Britten. Early in his career, the poet Auden had warned him that he had to choose between the distancing and decorum demanded by good taste and the need to plunge wholeheartedly into the creative act. In Auden's opinion, Britten was never able to arrive at that final and full acceptance of the stranger god. Many would disagree with Auden's judgment, yet it is clear that such a theme underlies Britten's final great work.

Introducing the gods Eros, Apollo, and Dionysus as subjects in this book is more than a literary conceit or metaphor on my part. They are real, living presences. Like it or not and call them what you will—archetypes, gods, or psychological dispositions—they are present in the lives of each one of us and our society. They represent the struggle between the opposing forces of order and sensuality in the mind of an artist or poet. They are at the heart of our own response to the natural world. In the creative act, the power of Eros is therefore confined within the boundaries of the specific, the particular, and the differentiated. Creativity is the meeting of Dionysus and Apollo, where poetic intoxication is tempered by the demands of structure, logic, and rationality.

A poetic insight may be immediate and striking. However, it is not an end in itself but merely the beginning of much hard work. A sudden flash of inspiration must find an appropriate poetic form. It must reanimate itself, using just the right words within a structure that clarifies, enhances, complements, and reflects each aspect of the whole.

The work of a stage actor must also strike a delicate balance between Apollo and Dionysus. There are times when an actor wants to take a risk, to make a leap of faith and entertain possession. But the same actor must also use a critical intelligence to stand back from the work at hand. In the moment, the actor may be inspired, but he or she must also know when to strip way, when to avoid excess, when to "lie truthfully." Unlike

Hamlet's advice to the players, the actor's goal is not to "hold a mirror up to nature" but rather to use skill, technique, sensitivity, and inspiration to create something living and truthful. The greatest plays and the greatest paintings are not mirrors to nature but are constructed with superb skill and truth.

In Francis Ford Coppola' film *Apocalypse Now*, the actor Martin Sheen performs a highly dramatic vignette in a bedroom at night. His character has been told to make a journey into the Vietnam interior to track down and execute a renegade colonel. That night, drunk, angry at being used, and in a rage against his fate, he paces his room. Sheen prepared for this scene, which was filmed in a continuous shot, by getting very drunk. Toward the end of the scene, when he sees his reflection in the mirror, he punched the glass so hard that he actually injured his hand.

The scene was gripping, yet I have always experienced a sense of unease in watching it, for its range of intensity seems out of keeping with the rest of the film. The actor was so drunk that he was on the edge of losing control of his performance. No longer was Sheen's body a fine instrument. Instead, he was incapable of remaining alert to excess, even to the point of injuring himself. During that scene, Sheen was embracing Dionysus and abandoning Apollo. This approach could never have worked on the stage; the measure of success Sheen achieved was due to the fact that the performance was filmed and could, in part, be created within the editing suite. To some extent, Sheen had sacrificed the exercise of his art by leaving the responsibility for critical discrimination in the hands of the director.

The painter Cézanne gave great importance to the "little sensations" he experienced before the motif. These sensations were the source of all his work, yet the act of painting itself was a long and sometimes painful process for him. His little sensations had to find their realization on the canvas in such a way that the entire structure of the painting was engineered as carefully as any bridge or skyscraper.

After a long period of frustration and false starts while trying to develop a difficult piece of new mathematics, Gules-Henri Poincaré had a sudden flash of insight as he lifted his foot to step onto a trolley car. But that flash was not sufficient in it-self. It had to find its particular form in the rigorous world of mathematical symbolism and be subject to the final court of appeal—mathematical proof.

Wagner's Ring Cycle begins with the opera *Das Rheingold*. For many years, the composer had struggled with the idea of a vast music drama that would be performed on successive nights. The task had haunted him since 1848, to such an ex-tent that at times he was struck down with a variety of neu-rotic illnesses. In the end, after retreating to Switzerland and beginning a course of cold baths, he began work on the li-bretto for the opera. This long poem about the world of gods and human was privately printed in 1853. By then, Wagner had all the elements of an opera cycle—except for the music it-self. On September 5 of that same year, he was staying at La Spezia on the Italian coast. Sick and half asleep, he became aware of the noises around him. "The rush and roar soon took musical shape within my brain as the chord of E flat," he re-called (quoted in M. Raeburn and A. Kendall, eds., *Heritage of Music*, vol. 3, *The Nineteenth-Century Legacy*, 1988).

This chord, repeated over and over again, gradually formed itself into the leitmotiv of the Rhine River. It ended up as the long, sustained sequence that begins the entire Ring. Like the river itself, the whole Ring Cycle flows from this chord. The chord was the seed for the many hours of music that follow, yet Wagner's cycle would never have flowered if the composer had not set himself the subsequent task of working out the en-tire musical structure of leitmotivs and developments.

During the year of the Great Plague, 1665, the young Isaac Newton (he was only twenty-three at the time) retired to his family's farm near Grantham in Lincolnshire and began to think about such topics as the color of light and the motion of the moon and planets. He derived the law of gravity in this pe-riod, but as late as 1684, he was still silent about his discovery. In part, he was directing his attention to the theory of color and his search for a "Universal Catholick Matter," but he also wished to present his theory in a complete and totally clear manner.

In 1684, he was pushed into action by the astronomer Edmond Halley, who informed Newton that astronomers were troubled about finding the exact law by which planets are attracted by the Sun. Newton replied that he had long ago solved the problem. Halley pressed him to publish his theory, but even then, Newton would present only a partial solution to the world. It was not until 1686, twenty-one years after his original insight, that the physicist published his *Principia* (*Philosophiae Naturalis Principia Mathematica*), which expounded, once and for all, his laws of motion and universal gravitation. By holding onto and, at the same time, withholding his discovery for so long, Newton was finally able not only to give an important mathematical result about the law of gravity but also to make it part of a vast logical and mathematical scheme, an approach to physics that would dominate science for the next two hundred years.

Describing the way he worked, the playwright Harold Pinter said, "The germ of my plays? I'll be as accurate as I can about that. I went into a room and saw one person standing up and one person sitting down, and a few weeks later I wrote *The Room*. I went into another room and saw two people sitting down, and a few years later I wrote *The Birthday Party*. I looked through a door into a third room, and saw two people standing up and I wrote *The Caretaker*" (from Pinter's conversation with Richard Finlater, reprinted in Harold Pinter, *Plays: Two*, 1977).

Another writer, Anthony Burgess, had a fleeting visual hallucination of a man sitting on the toilet writing. In this manner, the character of Enderby the poet was born. But in order to convert this sudden visitation into a series of books (*Inside Mr. Enderby, Enderby Outside, Enderby's End*, and *Enderby's Dark Lady*), Burgess had to involve himself with all the technical problems a writer faces—pace, structure, characterization, language, and so on.

Much is made of inspiration, yet the entry of Eros into the world demands the encounter with both Dionysus and Apollo, not only to provide the raw creative energy of inspiration but also to mold the material into its final form. That old joke about genius being 1 percent inspiration and 99 percent perspiration is more than a truism: Realizing an insight takes a great deal of hard work and discipline. There is a story told about the comedian Peter Cook. Encountering an old friend in a bar, Cook asked him what he was up to. "Oh, I'm writing a novel," the friend said. "Neither am I," replied Cook. Cook's point was that having an interesting new idea or exciting goal is one thing, but working it out in detail against the grain of the world is quite another.

The writer Beryl Bainbridge once talked to me about writer's block. This is a particularly painful situation in which the writer wants and needs to work but just can't put pen to paper (or hands to word processor). Similar conditions exist for artists and composers, and it is often taken for a temporary drying up of inspiration. This can sometimes be the case, and the artist can fall into a genuine depression. (The composer Michael Tippet told me that shortly after his ninetieth birthday, the inspiration of the gods of music left him—quite suddenly—and he realized that, although he had been musically active into extreme old age, he would never be able to compose again.)

But lack of inspiration probably accounts for only a small percentage of writer's blocks. Bainbridge believed that, in almost every other case, the block itself was about some technical point in writing. It was not Dionysus refusing to answer but Apollo making his demands. The insight for a piece of writing may be present, but for some reason, it is not being expressed in the correct way. Without knowing it consciously, the author realizes that something is wrong and, despite all that creative energy and an urge to get on with the story, a sixth sense is warning him or her not to continue. And so, the author has to wait, in a state that can range from rage and frustration to abject depression, until exactly the right means of expression is found.

The solution often revolves around a purely technical point. Maybe the story must be shifted from third to first person, from past to present tense; perhaps it needs to be told backwards or from several points of view. Charlotte Brontë's *Wuthering Heights* works so well because the story is told as a tale within a tale from the perspective of two commonplace characters, rather than through the eyes of the main protagonists, Heathcliff and Cathy Earnshaw. In this way, the extremes of passion in the story are given their greater range of dynamics by reference to the down-to-earth sensibilities of both narrators.

To take another example, the Western movie *Shane* had little impact when first shown to a sample audience. At this stage, the director went back to his editing suite and had the sudden inspiration of telling the story from the point of view of the hero's young son. The reedited version—exactly the same scenes but with a shifted viewpoint achieved through subtle editing—became one of the classics of the screen.

The act of progressive differentiation and structuring as a creative insight enters the world of matter, space, time, language, poetic form, or mathematical proof at both the individual and the collective levels. Copernicus had the insight that the earth revolves around the Sun, but it took decades for the scientists who followed him to work out the exact implications of his idea. It was only with Kepler's laws of planetary motion and Newton's three laws of motion and theory of universal gravitation that Copernicus's initial insight was finally made explicit within the field of physics.

At the end of the nineteenth century, physicists had worked out the formula describing the amount of energy radiated by a heated body. The result seemed nonsensical, since it predicted an infinite amount of energy being radiated. Then, in 1900, Max Planck proposed something truly radical: that radiation is not continuous but occurs in finite tiny packets called quanta. This insight changed twentieth-century physics, but its theoretical unfolding has taken an entire century and the work of much of the physics community. First, Einstein showed how these energy quanta interact with matter. Then, Neils Bohr tried to extend the theory to atoms. But it wasn't until 1925 that Werner Heisenberg finally developed quantum mechanics. Next, other physicists, including Paul Dirac and Enrico Fermi, worked out a quantum field theory that put matter and radiation on an equal footing. Even today, the story is not over.

Quantum theory has still to be reconciled with Einstein's relativity in a truly deep and satisfying way. A century after Planck's original insight, this final unfolding has yet to be achieved.

If the various examples given in the preceding passages suggest that the unfolding of creativity only occurs to great artists, musicians, writers, or scientists, this is not my intention. Creativity has a part to play in the life of each one of us. At some point in our childhood, we came across a book, talked to a special person, saw something that struck us intensely, or experienced something that was half dream and half vision. This special event, this voice that spoke to the heart, may have been carried inside us for years. If we were lucky and if chance smiled on us, then we found a way to unfold this creative germ within our own life—maybe through choosing just the right job, focusing on a hobby, encountering a partner, fostering a special set of friends, or building a place to live that became an expression of that creative moment.

Often, it's only when we look back at our lives and trace the paths we've taken that we finally realize the way some initial encounter has played itself out and established the pattern in our lives. In turn, we too now have the chance to become the creative germ for someone else, be it a child, a colleague, an organization, or a city park. Or perhaps we may simply, as Voltaire suggested in *Candide*, cultivate our own gardens.

To all this, I added a qualification—"if we're lucky"—because for some people, life remains such a struggle that this creative flowering never seems to happen. It could be that circumstances work against us or that we are just too fearful to take the plunge and risk a new sort of life. Yet it is never really too late. The creative germ never dies—it is merely dormant. There are times when life may seem dull and meaningless or when, for some unknown reason, we feel constantly depressed and unsatisfied. As Kierkegaard wrote: "These experiences may be painful but in truth they can be golden moments, for they remind us of what we had begun to forget, the unrealized creative potential for change that lies in each one us." Those moments of darkness are reminding us and prompting us to action. Creative chance is always possible, even in extreme old age.

Apollo and Dionysus also weave their way through contemporary physics. On the one hand, there is the Apollonian order of the laws of physics and the symmetry principles that govern the transformations of the elementary particles. On the other, there is the Dionysian chaos, when energy floods into a system to make it random and unpredictable. At a deeper level, order and chaos turn out to be not so much diametrically opposed forces as partners in a cosmic dance, a dance in which the one keeps changing into the other. Within the next chapter, we shall see how Dionysius and Apollo played out their creative roles in matter at the big bang birth of the cosmos, as well as during its subsequent structuring.

chapter 3

The Big Bang

I know there are readers in the world, as well as many other good people in it, who are no readers at all—who find themselves ill at ease, unless they are let into the whole secret from first to last, of every thing which concerns you.

—Laurence Sterne, *The Life and Opinions of Tristram Shandy, Gentleman* (1760)

It is no mere poetic metaphor to compare the birth of the universe with the work of an artist, or the way material forms unfold and evolve to the manner in which scientific insights turn into theories. Nor would it be inappropriate, for that matter, to compare such phenomena to the writing of a poem or the healing of strife within a community. When creativity enters the world, it must be vested in the lineaments of space, time, and matter. As it seeks to play within the realm of the manifest, it must inhabit form, order, and structure. So what can we say of matter and its origins? Just as in every artistic endeavor, the creativity of the universe must delimit and differentiate itself within the world of matter, space, time, galaxies, stars, planets, and even life itself.

I will be telling the tale of the origin of the universe in much the same way as Laurence Sterne's narrator, Tristram Shandy, tells the story of his own birth—that is, in fits and starts, evasions and excursions. This is only right and proper. Myths should always be told in this way, and the scientific story of the origin of the universe is a twentieth-century mythic artwork comparable to Dante's *Divina Commedia* or Homer's *Iliad*. To unfold it in a bland, linear sequence would be to betray its subtleties. And after all, I'm proposing that the story of the universe is not unlike the story of a painting, a novel, or a symphony. It's about the way the creative impulse is shaped, channeled, disciplined, and given form. It is about the alliance between Dionysus and Apollo, about intoxicated fecundity and the evolution of order and form.

If we're going to talk about origins, whether of the universe or of Mr. Shandy himself, allow me to indulge myself and go back in time to those years when, as a boy, I used to go fishing in Wales with my good friend Bill Mulligan. We'd fish until the light faded, and then, lying on our backs, we would look up at the night sky. The stars seemed cold and distant, yet as I gazed up at them, there always came a curious sense of connection, as if I and the stars were bonded directly. I felt that, at any moment, my body would leave the surface of the earth and fly up into the farthest reaches of the universe.

Later, when I went to university, I learned that my sense of a connection to distant stars was not all that crazy. It is true that the stars we see are hundreds and thousands of light-years away. Yet for that photon of light that leaves the surface of the star and journeys to my eye, its *proper time*—the time lapse experienced by the photon itself, if that makes any sense—is zero. For the photon, the distance it has traveled is likewise zero.

This may sound absurd, but everyone knows that time slows down and distances become shorter as you approach the speed of light. At that speed, the speed of a photon, time and distance shrink to zero. Put another way, the star and I are essentially in contact, joined by a photon that moved between us in no time and no space at all.

That scientific revelation of my direct contact with all the matter of the visible universe seemed like a miracle to me at the time. But now, as I type this text, I am aware of another miracle. I look down at my hands and ask: "I am in contact with the stars, but where did they come from? Where do I come from? In what manner did my hands, my body, and all the matter out of which I am built come into existence?"

The act of just being, of writing this book, of looking down at my hands, or of getting up each morning is the culmination of a series of miracles that began with the origin of the cosmos itself. Each is an expression, an evolved manifestation, of the essential creativity of the cosmos.

My questions cannot be answered by reference to Earth alone. To understand the answers, we must all journey to the stars, travel though vast tracts of interstellar dust, enter the core of suns, and journey back to the origin of time itself. We must trace the story of how an initial creative impulse enters time and, in manifesting itself, limits itself though the laws and symmetries of nature.

The Big Bang

Scientists believe that all that exists, the entire cosmos of space, time, energy and matter, began in a single event—the big bang. This statement needs a certain degree of qualification. By *scientists*, I mean the general consensus of the scientific community. And although there are some technical glitches and arbitrary assumptions in the big bang theory, it is accepted by all but a few maverick physicists and cosmologists who have proposed alternative theories.[1]

A further qualification concerns the term *all that exists*. Scientists are often jokingly referred to by philosophers as "naive realists." For scientists, what is real is what can be touched, held, observed, and manipulated. "The beginning of everything" means the origin of time and space, matter and energy. They don't really feel comfortable speculating beyond that point.

---

[1]Some theories suggest that our universe is a rebound from a previous one that collapsed. So in the distant future, our own universe will collapse right down to an incredibly small size and then bounce back again as a totally new universe. This is not strictly an alternative to the big bang theory but rather a gloss to the established approach.

So let us begin. We're going to start not at the ground zero of time but 1/100 of a second after the universe had begun because from this point onward, scientists are pretty clear as to what happened. It is that tiny interval before, a time as short at the click of a camera shutter, that is more mysterious.

What was happening 1/100 of a second after the universe had come into existence? At that point, the universe was a fireball with a temperature of 100,000,000,000 (or 10^{11}) degrees—far hotter than the center of the hottest star and more intense than a hydrogen bomb explosion. What's more, the entire universe was packed into a space of around 4 light-years, the present distance from us to the nearest star.

Already at 1/100 of a second, an enormous amount had been happening. Indeed, 1/100 of a second is a very long time when it comes to the birth of the cosmos, and in that span, the universe had already passed through several different eras. Much earlier, there had been a period of explosively rapid expansion, but now the universe was slowing down, doubling in size only every 2/100 of a second.

It was not like the universe we know today. There were no galaxies or stars. Rather, there was just an intensely hot soup consisting of light, electrons, and neutrinos packed together with a density several billion times greater than that of water and encountering each other in constant bursts of creation and annihilation.

One-tenth of a second had to pass before the temperature dropped sufficiently (to a few ten thousand million degrees) for the heavier particles to form out of the collisions of electrons and positrons (antielectrons). In this period were born the neutrons and protons that eventually formed the cores of all the atoms in the universe—the same matter of which we are made! At this stage, the expansion slowed down to the point where the universe only doubled in size every 2/10 of a second.

After a full second had elapsed, the entire universe was still a thousand times hotter than the center of our Sun and much too hot for the protons and neutrons that were now being produced to combine and form the nuclei of the first atoms. The universe would have to wait for a full three minutes until it had expanded and cooled down sufficiently for the first nuclei of helium to form (hydrogen being the first and most abundant element in the cosmos, helium being the next in line). At this point, three minutes after the big bang, things were still too hot for actual atoms to form, but all the nuclei of all the hydrogen and helium that make up our universe had already been born.

From this point on, the universe would continue to expand at a slower and slower rate and, at the same time, cool down. Eventually, the universe cooled to the point at which the first stars formed and the heavier elements began to be synthesized in the core of stars. Finally, the first molecules of protolife would form.

Thus far, we have covered only the first few minutes of creation. Much more of this story has yet to be told, by going both forward and backward in time. But now, like Tristram Shandy, I will pause and follow another train of thought. The birth of the universe occurred in an explosion far greater than that caused by any hydrogen bomb, involving both temperatures vastly hotter than those inside any star and energies beyond those that any elementary particle accelerator on Earth will ever produce. Compared to this explosion, the world's entire stock of hydrogen bombs are mere fireworks.

We all know about Einstein's magic formula, $E = mc^2$, which tells us that energy and matter are interconvertible. With all that energy present at the birth of the universe, new elementary particles must have been endlessly flickering in and out of existence. Our experience here on Earth has made us familiar with nature's proclivity. We have come to expect species upon species. From the depths of the sea to deserts, jungles, and mountaintops, life exists in a host of unexpected and bizarre forms. But if this is true for species, why would it not also be true for elementary particles? Why aren't there millions upon millions of totally different particles, exhibiting curious and unexpected properties and combining in endless ways to form hosts of bizarre chemical elements?

We know this is not the case. But what is stopping anything that we could imagine from actually taking place? If the cosmos was born in an incredible burst of creative energy, then why is our universe not filled with square planets, green suns, and polka-dot galaxies? Why indeed does the universe appear so uniform wherever we look?

The answer lies in nature's fundamental principles of symmetry. Put another way, when creativity seeks to display itself in the manifest world, it must take on the limitations of structure and form, much as a poem must. With poetry, the confines are denoted by the poem's form—for example, the strict sonnet form of Milton, Wordsworth, or Shakespeare; in the case of matter, symmetry principles and conservation laws of physics apply. And once again, all that exists is brought about through the powers of Apollo and Dionysus.

Physicists have a deep faith in the uniformity of space— that is, in a sort of ideal space before the presence of matter and energy. The laws of physics are founded on the principle that space is isotropic (in whatever direction you look, space is the same) and homogeneous (as you move from place to place, space does not change). There is also the symmetry of time, which means that the laws must be the same from past to future as from future to past. The symmetry between the positive and negative aspects of electrical charges governs the electrical attractions and repulsions between elementary particles. Other symmetries concern the strong nuclear force between elementary particles and what is known as the weak force.

Whenever such fundamental symmetries exist, scientists know that some property must also be conserved—that is, some particular form in the material world will be unchanged no matter what reactions and transformations occur. The very fact that space is symmetrical means, for example, that momentum must be conserved in all possible reactions—from the collisions of billiard balls to the annihilation of electrons and positrons into light. (The homogeneity of space implies that linear momentum must be conserved. Its isotropic symmetry means that angular momentum is conserved.)

The conservation of angular momentum, which follows from

the isotropic symmetry of space, rules the dynamics whereby rotating galaxies and planets are formed. It is the reason that a skater speeds up as she twirls. You can test the conservation of angular momentum for yourself by sitting in a swivel chair and holding two heavy books at arm's length. Give the chair a twist, and quickly bring the books into your sides. Notice how much faster you spin. This is because angular momentum is conserved—a compact body must rotate much faster in order to have the same angular momentum as a more slowly rotating extended body.

By adhering to this principle, Wolfgang Pauli gained the Nobel Prize in physics. Physicists had long observed a form of radioactivity called beta decay, which is associated with the emission of electrons. Many radioactive isotopes and an elementary particle called the neutron exhibit beta decay. A proton's life is much longer than that of the universe, but the neutron happens to be a tiny fraction heavier than the proton. This means that it can gain greater stability by losing this extra mass. It does so by shooting off an electron and decaying into a proton. The neutron is electrically neutral, so in emitting a negative electron, it can get rid of excess energy and, at the same time, exactly balance the positive charge created on the proton. (The conservation of electrical charge is demanded by another basic symmetry—that between positive and negative changes.)

When Pauli looked into the concept of the neutron's beta decay, he realized that something was deeply wrong. The neutron has angular momentum (called "spin 1/2" in quantum mechanics). So does the proton and the electron.

This means that

$$N = p + e$$

$$(+1/2) = (+1/2) + (-1/2)$$

The left-hand side of the equation has spin (+1/2), the right hand side has spin (+1/2) + (-1/2), which equals zero. Somewhere, a unit of angular momentum has vanished.

No matter how hard physicists looked, they could see nothing else going on in the decay but the emission of an electron and the resulting creation of a proton. Since nothing else was registered on their particle detectors, they began to speculate that maybe angular momentum was not conserved in the quantum world. Pauli, however, insisted that these symmetries and conservation laws had to operate throughout the entire scale of the universe. It they didn't, then the universe would not make sense to him. The only way this could happen was for an invisible third particle to be involved. It had to be a particle with no electrical charge and no mass so as to be undetectable; all it could have was a spin—one half (1/2).

Pauli made his claim in 1933, but it wasn't until 1956 that scientists finally found evidence for this "neutrino," v.

$$N = p + e + v$$

$$(+1/2) = (+1/2) + (-1/2) + (+1/2)$$

Now the total spins on both sides add up to +1/2, and angular momentum is conserved.

This is no idle example. After the first 1/100 of a second of the universe, all that was really around were electrons, neutrinos, and light. (A tiny number of neutrons and protons were present, and over the subsequent time, they would grow in number but always be confined by the restrictions of angular momentum conservation.)

It turns out that the entire cosmos is based on a handful of conservation laws, some of which were mentioned earlier (time, charge, and momentum; the total sum of matter and energy; and the total sum of the light elementary particles [leptons] and heavy particles [baryons]). Additional symmetries and conservation laws are important (such as conservation of parity), but they do not hold under all possible conditions.

Thanks to these fundamental symmetries, the big bang is not the tuning up of a cosmic orchestra; neither is its subsequent development a free-form improvisation. Rather, it is a symphony that, though containing elements of freedom, must adhere to strict musical forms.

An analogy can be made to the canon in music. A canon is a particularly rigorous form of fugue in which some aspect of the theme is always sounding. Bach's canons appear endlessly creative, but they are actually constructed out of a single theme, which is transposed into different keys, played faster or slower, inverted into a mirror image, and even played backwards. Thus, the music is filled with endless possibilities, yet it is based on fundamental symmetries and invariances. No matter how the music sounds, the exact form of the theme must be present in one of its musically symmetrical transformations.

Likewise, during the big bang, matter could undergo endless transformations and space and time could twist and curve, but the fundamental symmetries always had to be preserved. This is why, as creativity expands into the material universe, it is confined by laws and principles. In its Dionysian origins, the physical manifestation of the universe was restrained by Apollonian order.

Now let us pick up the story of our universe's birth again. At our starting point of 1/100 of a second, the conservation laws were confining matter to appear in the form of electrons and positrons, neutrinos and antineutrinos. For every thousand million of these, there would have been one proton and one neutrino. All were constantly transforming into each other, not in a willy-nilly way but so as to preserve the conservation laws of charge, momentum, mass-energy, and so on.

As the universe continued to cool and expand, the speed of and the energy carried by these particles decreased. Soon, electrons and positrons were annihilating faster than they were being created, and the number of neutrons and protons was beginning to increase. At the same time, the pull of gravity was slowing expansion. After the end of the first half hour, all the helium and hydrogen of the universe had finally been produced, and the universe was doubling in size every hour or so. Its temperature was now measured in ten of millions of degrees.

Things now happened a little more slowly. It would take a hundred thousand years of continued expansion before the universe was cool enough for the first hydrogen and helium atoms to form. Up to that time, they had been coming together and fusing but were immediately ripped apart again by the intense energetic light that bathed the cosmos.

A few million years later, the universe had stretched out and cooled. Its matter was now much farther apart. Here and there, the force of gravity was causing this matter to lump together. Some of these clumps were very slowly rotating. As they condensed, like the skater pulling in her arms, the conservation of angular momentum caused the mass of matter to speed up and form the first rotating galaxies. This distant past was also the era of star formation and the period in which the elements that make up our bodies were synthesized.

At the center of many of these galaxies were black holes, regions of space-time curved by gravity to such an extent that the very fabric of space-time had ripped. While nothing, neither matter nor light, could escape from the black holes, their vast gravitational attraction pulled in everything around them. In effect, they acted like cosmic vacuum cleaners. As the interstellar gas of the galaxy moved toward the black holes, it accelerated and emitted radiation. The result was something on a much grander scale than a mere star. The light that was produced was of such intensity as to make the regions around these black holes by far the brightest objects in the early universe. Today, we know them as quasars. They can be seen in those parts of the universe that are most distant from us because the farther away we look, the farther back into the past we are looking.

There are no quasars around today. Those that astronomers "see" are actually signals quasars sent out billions of years ago. They have long since died out and become relatively inactive black holes, for all the gas in that region of space has been sucked away.

Synthesis of Elements

When I was a young boy, even before I could read, I loved to look though the pages of books such as *The Marvels and Mysteries of Science*. They told, in part, the story of our origin, the way we were all born out of the Sun. A wandering star was supposed to have approached our Sun, drawing out a great cigar of solar atmosphere that later condensed into the planets. Our Earth and own bodies, the story went, were created out of the matter of the star that heats and lights our days. The problem with this account is that we now know it is badly wrong.

The true story is more amazing. We were never born out of the Sun but are the descendants of great element factories deep within the cores of suns that are long, long dead. Our Sun is relatively young, a member of the third generation of stars that have inhabited our universe. The matter within our bodies is far more ancient than the Sun itself.

After the first few million years of the universe, the hydrogen gas that makes up most of the matter of our cosmos began to clump together. As these clumps became denser and more compact, more gas became attracted inward, and, like Newton's apple, it began to fall toward the center of what was destined to become a star.

As the atoms of gas fell inward, they accelerated under the star's gravitational attraction, colliding with each other and causing the gas to heat up. In this early stage of star formation, gravitational energy was being converted into heat energy. Eventually, the gas began to glow with heat, its temperature increasing all the time until, at ten million degrees, the first nuclear reactions began.

There is an electrical repulsion between the protons that make up the nuclei of hydrogen. These protons have to be hot enough (which means traveling really fast) before they can get close enough to react. (This is the same sort of reaction scientists are currently trying to achieve on Earth using hydrogen plasma to produce controlled thermonuclear fusion.)

In this first stage of a star's life, a series of nuclear reactions "burned" its hydrogen to produce helium began. Through a series of collisions, the nuclei of four hydrogen atoms fused together to produce the nucleus of helium. Since four hydrogen atoms weigh slightly more than the resultant helium atom, a tiny amount of mass was released in the form of energy during the fusion reaction. This is the same energy that is given out in the hydrogen bomb and that powers our own Sun.

These nuclear reactions continued for several billion years until the star had burned a significant percentage of its hydrogen. The core now contained a high concentration of helium, and hydrogen burning was only possible in the star's outer shell. Up to now, the high temperature of the core—which tended to cause the gas to expand—had opposed the inward force of gravity. But as hydrogen burning died down in the inside of the star, the temperature dropped and the core collapsed inward. As with the star's original creation, this collapse now caused the helium nuclei to accelerate inward and undergo ever more rapid collisions. The result was that the core heated up again as more gravitational energy converted into heat energy. At around 120 million degrees, helium nuclei were moving so fast that they could overcome their mutual electrical repulsion. Consequently, these nuclei now fused together in a new series of nuclear reactions.

The previously inert helium was now burning and producing heat. A new era in the star's life had commenced. The by-product of this new thermonuclear reaction was beryllium. The beryllium was so unstable that it decayed almost as soon as it was formed, but now and again before this decay happened, a beryllium nucleus was hit by another helium nucleus. As a result, a nucleus of carbon was formed. A further fusion reaction produced the element oxygen. Then, at even higher temperatures, carbon and oxygen began to react to produce the elements magnesium, sodium, silicon, and sulfur. At an even higher temperature, silicon nuclei began to "burn," producing an ash of chromium, manganese, iron, cobalt, and nickel.

At each stage, higher and higher temperatures were required to enable nuclei to move fast enough and get close enough to react. The final barrier was reached with the element iron. No amount of temperature increase and no nuclear burning would ever create the many elements that lie above iron in the periodic table.

But our Earth does contain heavier elements than iron, among them lead and uranium. Where did they come from if not from the heart of stars? The nuclear fusion that occurs when two smaller nuclei meet at ultrafast speeds won't take place with the heavier elements. What happens instead is that neutrons—which are always flying around in the center of stars—collide with a nucleus and bind to it. This increases the nuclear mass by one unit. Generally, this means that the new nucleus is unstable, for stability only occurs when the number of protons and neutrons is about the same; if there are too many neutrons, the nucleus is radioactive. To achieve stability, the nucleus uses beta decay and shoots out an electron. Since a negative charge has been thrown away, the neutron turns into a proton, and the resulting nucleus has moved up one step in the periodic table to become a new, heavier element. In this way, moving step by step though the absorption of neutrons, the heavier elements were created inside a star.

However, nuclear burning cannot continue indefinitely. The amount of fuel available is limited, and the star is doomed to die in one of several ways. It may simply burn out into an icy ball, around the size of our Earth but with a density fifty thousand times greater. It may continue to collapse right down to a neutron star, only about twenty miles across but so dense that it acts like the nucleus of a single atom. Or, if it was heavy enough to begin with, its end point will be a black hole that sucks matter and light in toward its space-time singularity.

In some cases, the star explodes dramatically as it reaches its final stage of life. It thereby becomes a supernova, and for a time, it is the brightest object in its galaxy as it spews out its nuclear matter though space. In this way, empty space gradually becomes filled with a tiny percentage of heavier elements, some of which drift as clouds of interstellar dust, others that are destined to form the cores of a new generation of stars. (Incidentally, element building can continue even in the cold depths of space. Space is filled with cosmic rays, fast-moving elementary particles, and light nuclei that have been accelerated by the magnetic fields that twist through galaxies. When they collide with the nuclei spewed out from supernovas, they are able to fuse to form heavier elements.)

Periodic Table

In this way, the elements were built up to form the components of our Earth and eventually our bodies. It turns out that there are around ninety different elements in the universe— and over a hundred if you count the ones created in the laboratory. With such a diversity of species, all of which can combine to form molecules, we would expect Earth to be a bizarre Aladdin's cave. Just think of all the possible combinations when ninety elements are permuted in all possible ways to form different molecules. It would be natural to assume that the result would be a world without any apparent order, one in which each particle of dust is totally unlike every other. Yet Earth is a pretty regular place, where dust is dust and water is water. How can such order have emerged from the potentially astronomic combination of ninety different building blocks?

The answer was given in the middle of the nineteenth century by the Russian chemist Dmitry Ivanovich Mendeleyev, who showed that rather than being all totally different, the chemical elements form just a few families whose members react in similar ways and form similar sorts of molecules. Again, Apollo had created order out of Dionysian proclivity. This order came about in the following way.

Molecules are formed from atoms. Atoms consist of a cloud of electrons surrounding a dense core, or nucleus. To preserve electrical neutrality, the number of (electrically negative) electrons in the cloud must be exactly equal to the number of (positive) protons in the nucleus.

The nucleus itself contains a roughly equal number of positive protons and electrically neutral neutrons. If there are too many neutrons, the nucleus is unstable and radioactive. (As we have already seen, one way it can achieve stability is though beta decay—the emission of an electron that, in effect, transforms one of the excess neutrons into a proton. Now, the nucleus will have an additional electrical charge, and the electron cloud will have one extra electron.)

The number of protons in the nucleus is called the atomic number, a useful tag whereby scientists label an atom. Because of electrical neutrality, this atomic number (the number of protons) is also equal to the number of electrons in the outer cloud. Hydrogen has an atomic number of 1—a single electron surrounding the proton core. Helium has an atomic number of 2—an inner core of two protons and two neutrons and an outer cloud of two electrons. Next in line is lithium, atomic number 3, with three protons and three neutrons in its core and an outer cloud of three electrons. And so it goes on up to uranium, which has an atomic number of 92—ninety-two protons in its core and ninety-two electrons orbiting the nucleus. (Elements that are heavier than uranium, such as americium, californium, fermium, and einsteinium, can be produced in the laboratory but are not found in nature!) Isotopes are elements that have the same atomic number—the same number of electrons in the outer shell and the same number of protons in the core—but have different numbers of neutrons. There are three isotopes of hydrogen: normal hydrogen, with one proton; deuterium—the isotope of hydrogen found in heavy water—with one proton and one neutron; and radioactive tritium, with one proton and two neutrons.

By carefully tabulating the properties and chemical reactions of all the elements, Mendeleyev realized that they fall quite naturally into family groups. The family photograph album of the elements is called the periodic table. It begins with the elements hydrogen, helium, lithium, beryllium, boron, and so on and groups the family members in columns under each of these elements. When Mendeleyev had finished his table, he discovered gaps and reasoned that these gaps did not represent a flaw in his classification system but had to stand for as yet undiscovered elements. By comparing the missing elements with other members of the same family, Mendeleyev was able to predict their properties and chemical reactions. By the end of the nineteenth century, several of the missing elements had been discovered, and they fitted perfectly into their predicted positions.

Mendeleyev had developed his table by comparing and contrasting the chemical reactions of the various elements. It was only in the following century, with Neils Bohr's development of the theory of atoms, that the underlying reason for the similarities among elements became apparent.

As we saw in the previous chapter, Bohr's picture of the atom was a mixture of quantum and classical ideas, and it was replaced by the quantum theory of Heisenberg and Schrödinger. But it does provide a good pictorial approach for describing atoms. And, though it is a considerable oversimplification, its explanation for the existence of the periodic table is basically correct.

Bohr showed that the electron cloud that surrounds an atom likes to group itself into a series of stable shells. Helium's pair of electrons forms a very stable shell indeed. This is why helium is chemically inert. The same degree of stability occurs for neon, with an atomic number of 10. Neon's ten electrons form two stable inner shells (with two electrons in each shell) and a stable outer shell of six electrons.

Next in the family comes argon, whose eighteen electrons also form stable, closed shells. All the gases (neon, argon, krypton, xenon, and radon) can be listed in the periodic table in a column below neon to form the family called the noble gases. These elements are chemically inert to a very high degree.

What about the element that comes next in line after helium? Lithium, with an atomic number of 3, has a lone electron lying outside the stable inner core of two electrons. It is an electron in a chemical singles' bar, just waiting to get picked up and find stability in a relationship with other electrons!

Other members of this family, called the alkali metals, also have lone electrons lying outside stable inner shells. They include sodium, potassium, rubidium, cesium, and francium. All are chemically reactive: Drop a piece of sodium metal in water, and it reacts so violently with the water that it bursts into flame; rub a piece of potassium on a rough surface, and that small degree of friction causes it to burst into flame (which is why potassium was used to make the first matches).

The alkali metals differ from the noble gases by having one additional lone electron in an outer shell. At the other end of the scale are those elements that lack one electron to complete their outer shell. Fluorine has an atomic number of 7. With just one extra electron, it could achieve the stability of neon (atomic number 8). Instead, however, it is doomed to wander the chemical singles' bar in search of an element that has a single electron to spare.

This family is called the halogens, and it includes chlorine, bromine, and iodine. Just think what happens when highly reactive sodium, with a lone extra electron, meets the gas chlorine, which has always known that something was missing in its life. The two combine chemically, never to be parted again, and the result is a molecule of sodium chloride, or common salt, which is very stable chemically.

The periodic table has several other families of elements. The elements in each column have similar chemical reactions. Thus, if you are on a low-sodium diet, you can simply substitute potassium chloride for sodium chloride. It lies in the same column as sodium and therefore reacts in a similar way and has a similar salty taste.

One family begins with carbon, and its nearest relative is silicon. Both carbon and silicon share the ability to build up enormously complicated chains of molecules. Life on Earth is based on this ability of carbon to form complex molecules. Some scientists have speculated that silicon-based life could exist on other planets. (In fact, as silicon-based information technology increases in computing power, there could one day be a symbiosis between carbon and silicon biology to produce new species of intelligent organisms.)

Mendeleyev's periodic table and its confirmation by the quantum theory of the atom again shows how order and regularity always temper the basic creativity of the material universe. If the universe is a work of art, then it is a work that demands both discipline and structure and a passionate energy of creativity.

The Birth of Earth

Many billions of years ago, the chemical elements that had been synthesized in the heart of long-dead stars spewed out into a gas that drifted through our galaxy. About four and a half billion years ago, a mass of gas part of the way out in one arm of the Milky Way galaxy began to clump together, increasing its speed of rotation as it did. Out of these gases, a star and planets began to condense. The result was our own solar system. Condensation of the Sun under the force of gravity heated it to such a point that thermonuclear fusion began, with hydrogen burning to produce helium. Each of the planets also heated up as it condensed. Jupiter heated almost to the point at which nuclear fusion occurred in its hydrogen. If it had been just a little more massive, it would have been transformed into a second sun orbiting our own.

In these early days, Earth was a molten mass whose surface gradually cooled. Its atmosphere was highly poisonous to what we today know as life. It consisted of ammonia, methane, and hydrogen sulfide. Earth was a planet racked by electrical storms and radiation from its young sun. In addition, the radioactivity of many of its rocks was more intense than today. Even long after this period, when the earth had cooled and its crust had formed, isotopes of uranium, such as U-235, were still abundant enough for spontaneous fission reactions to occur, just as they do in nuclear power plants today. (Remnants of such a naturally occurring fission reactor have been discovered in Africa.) Today, four billion years later, these isotopes have decayed to such an extent that their concentrations are much lower.

Earth was also subject to bombardment by meteors. This was an early time in which the dynamics of the solar system were still relatively unstable. These meteors brought with them not only metals such as iron, nickel, and cobalt but also carbon and water, the raw materials out of which life could be built.

Gradually, iron collected into a molten core, leaving a solidifying, iron-rich mantle of lighter materials. In time, the surface of the earth cooled. Continents formed, and oceans and mountain ranges appeared. Earth's atmosphere was still rich in methane, ammonia, and hydrogen sulfide.

Time passed, and Earth cooled even further. If intelligent beings had been monitoring our Earth from outer space, after a long time they would have suddenly noticed a dramatic change that could not be explained on the basis of everyday chemical reactions. Earth's atmosphere began to transform. Ammonia, methane, and hydrogen sulfide began to disappear and be replaced by oxygen and nitrogen. Only one explanation was possible: Life had appeared on Earth.

Life is prolific. We do not know its end point or goal. But we human beings have a tendency to see things from our own point of view. Looking back at the history of the universe and the evolution of life, we are sometimes seduced into believing that the whole process was a pathway inevitably leading to the development of intelligence and ourselves. However, for nature itself, the human race may be rather insignificant—far less important than the cooperative colonies of microorganisms, for example, that help maintain the salinity of the oceans and the chemical composition of the atmosphere.

But we are here, and we're probably not alone in the universe. The universe has produced us, and we are able to observe and reflect on that universe. Above all, our human characteristic is to ask questions, some of which have mythic proportions: Where did we come from? And how did it all begin? All myths and religions attempt to answer these questions.

Up to now in this chapter, we have avoided these pregnant questions. We only went back as far as the first 1/100 of a second after the big bang. Now we must turn the clock back even further and try to find out where we all came from.

Back to the Origin

We began the story when the universe was 1/100 of a second old. Already, a great deal had happened, and our cosmos had passed though several eras of growth. To continue the story, we must tell it in reverse and cover shorter and shorter intervals of time.

Closer in time to the instant of the big bang, the universe became smaller, denser, and hotter. This means that the elementary particles were moving so fast that the energy released in their collisions created heavier elementary particles. The temperature of one hundred thousand million degrees (10^{11} degrees) occurred in the era of the heavy elementary particles—protons, neutrons, and mesons. Even earlier, things were so hot that these heavy particles broke down into their constituent quarks.

At temperatures so high that twelve zeros must be used (10^{12} degrees), the universe consisted of photons of light, quarks, and leptons (electrons and neutrinos). But if we are to go back any earlier in time, we first have to understand one of the cornerstones of modern physics—symmetry breaking.

Symmetry Breaking

As I emphasized before, despite its rational face, science is based upon a belief in the uniformity of nature. Scientists have a strong faith that the universe makes sense. It is not chaotic and arbitrary, they believe, but can be understood by the human mind. For their research to have any meaning, the universe must have been built on rational laws and principles. Moreover, these fundamental laws should be mathematically elegant and simply stated.

This is quite an anthropocentric approach but one in keeping with physicists' tendency to believe that when they ask fundamental questions, they are probing the mind of God. Look back to those conversations in the early decades of the twentieth century between Bohr and Einstein and notice how often "the Good Lord" is invoked! Of course, when they happen to cast God's mind in the image of their own, scientists are being pretty unsophisticated compared to contemporary theologians!

The idea of a creator basing the universe on the simplest and most elegant principles of all is a beautiful and seductive notion. Yet physics may eventually reveal that nature does not work this way at all. It could be that the desire for simplicity and elegance is simply the projection of a certain type of consciousness that grew out of the Greek and Judeo-Christian worldview.

Writing down the simplest possible laws of nature implies a totally symmetrical and democratic universe. In such a universe, all the forces of nature—gravitational, electrical, and nuclear—are equivalent. All the elementary particles are on an equal footing, differing only in their spins, electrical charges, and so on but having equal masses. It is a universe in which things are identical in whatever direction we look, a universe in which it does not matter if time goes forward or backward.

The only fly in the ointment is that this is not the universe we happen to see around us. The elementary particles are not democratic; the proton, for example, is around two thousand times heavier than the electron. Neither are the forces of nature equal in strength. A comb that has been run though your hair will pick up scraps of paper from your desk. In other words, the tiny electrostatic charge on the comb is sufficient to overcome the gravitational pull of the entire earth!

How can the seductive vision of elegant simplicity be saved and, at the same time, the universe we live in be accepted? How did it come about that a universe based on democratic and totally symmetrical laws ended up with such a degree of grain and baroque decoration? The answer, scientists propose, is something called symmetry breaking. Another way of putting it is that the universe represents a partial Dionysian victory over the power of Apollo. The particular and individual must also exist within the world of the universal and general.

Think of the universe as a ball poised at the tip of a conical hill. All around the ball lies a perfectly symmetrical landscape. From the perspective of the ball, the world is the same in every direction. This is a metaphor for the vision of nature determined by totally symmetrical, democratic laws. But this ball-universe has to evolve. It can't stay poised forever in a symmetrical initial state, and so it begins to roll downhill.

The instant before it began to move, all potential paths were equivalent. Once it begins to roll, it has selected, by chance, one path from an infinite number of possibilities. And so, as it runs downhill, the initial symmetry of the situation is broken. Space is no longer the same everywhere because an individual path, a particular direction, has been singled out by the actuality of events. This is what scientists mean by symmetry breaking. The laws of nature retain their symmetry, but actual, physical, individual events no longer have such a high degree of symmetry. (At a deeper level in the theory, physicists are able to show that the original symmetry is still present but hidden within the dynamics of particular situations.)

Take an example. The laws of magnetism are written in the most symmetrical form possible. They indicate that each direction in space is equivalent. But once an actual magnetic field is present, once a real bar magnet is brought into the picture, then a particular direction is singled out in space. In the absence of a magnetic field, a compass has nowhere to point, for magnetism is potentially totally symmetrical. But once a real magnetic field appears, one born out of a totally symmetrical equation, the compass points in a definite direction, and the initial isotropy of space is broken.

Just as an actual magnetic field breaks the isotropic symmetry of space, so too does a crystal lattice break the homogeneity of space. The fundamental laws of physics have such symmetry that every point in space should be exactly the same as every other. But in a crystal lattice, such as that of common salt (which is a collection of cubes), certain positions, or planes, are singled out. The lattice certainly has a high degree of symmetry but not as high as a homogeneous space. Again, an initial symmetry has been broken.

Symmetry breaking also occurs when water freezes into ice. At room temperature, water molecules are moving around and colliding at random so that, on average, every point in the water is the same as every other point. But as soon as the ice freezes, a lattice is formed, and certain planes in space are singled out.

What applies to magnetism and the freezing of water is equally true of elementary particles and nuclear forces. The fundamental nature of the universe is supposed to be described by mathematically elegant equations. These laws are a little like Platonic ideals. They are pure forms that exist in an ideal world, whereas the actuality of the universe we happen to live in is of a much lower symmetry. However, it turns out that the closer we move to the initial moment of the universe, the closer we approach the aboriginal, ideal symmetrical state.

This initial, highly symmetrical state has much more energy than the symmetry-broken states, in which particles have different masses and the forces of nature have different strengths. Cooling the universe means moving away from its highly symmetrical origin through a progressive breaking of symmetries. Apollo may have dreamed up an ideal universe, but Dionysus thereafter holds sway over the manifest world.

Above 3×10^{15} degrees, the electrical and the weak nuclear forces were united into a single electroweak force. Until the universe cooled below this temperature, the force that governed beta decay had exactly the same strength as the electromagnetic force. Earlier in time and at even higher temperatures, this electroweak force was unified with the strong nuclear force. And, as the forces of nature moved toward the unity of a single force, the masses of the elementary particles became equal.

Physicists are hazy about what was going on even further back in time, at around 10^{-43} second from the big bang. There are good arguments to suggest that when the temperature of the universe was 10^{32} degrees, the force of gravity was unified with the other forces of nature to produce just a single force and total democracy of all the elementary particles.

But in a universe so close to the big bang origin, the very notions of time, size, and particle cease to make sense. To begin with, the "size" of the universe was, in fact, smaller than the "size" of an elementary particle! So what does it mean to speak of particles being created and annihilated? And in such a situation, what is the meaning of time and the succession of events?

Despite a century of effort, physicists have been unable to come up with a clear and consistent theory that unifies relativity and quantum theory. But as their theories approach that primordial moment, the origin of everything, they nevertheless believe that the universe must have existed in a unified state governed by a single fundamental law. Although there is no truly satisfying answer to questions about this early period of the universe, there have been a number of speculations. One of these involves the theory of superstrings. This theory posits that, at ultrahigh energies and incredibly tiny dimensions, the world is not made out of elementary particles at all but of patterns of rotations and vibrations—the quantum states of extended objects called superstrings. What's more, these superstrings do not exist in our own space and time but in a space of fifteen dimensions!

According to this theory, the universe began as fifteen incredibly tiny dimensions containing vibrating strings. But even this statement is not quite true, for in such an era, the geometry of the universe could not be separated from the superstrings themselves. Rather than strings being "in space," our notion of what is a particle and what is space would have to emerge out of such a theory. (The idea of superstrings is ingenious but, as it stands at the moment, not really satisfactory. The problem is that the theory contains a number of arbitrary elements. In fact, there are many superstring theories and no really good way of deciding between them.)

Let us accept the picture of a fifteen-dimensional universe for a moment. For some reason, this universe began to expand in such a way that four of these dimensions (our three dimensions of space and one of time) started to unroll and expand, while the others remain compact.

Now let's allow the clock to run forward from the era of fifteen-dimensional superstrings. As the three dimensions of space began to unroll, the vibrating patterns of energies of the superstrings began to look, though a grainier microscope, like a highly symmetrical world of democratic particles and a single unified force within a three-dimensional space that was expanding in time.

This universe continued to expand and cool. Just like cooling water that forms into ice, the forces of nature began to crystallize out. First gravity separated out, then the strong nuclear force parted company from the electroweak force, and finally the electrical and weak nuclear force separated. At each stage, symmetries were being broken and the masses of the elementary particles were differentiating. But along this road of symmetry breaking, something curious occurred, something that is rather like the supercooling of water—the supercooling of an entire universe!

Pure water freezes at 0°C. But if this water is superpure, with no traces of dust particles, then it is possible to take the temperature even lower so that water becomes supercooled to several degrees below zero. This supercooled water exists in a highly unstable state. It is like the ball poised on top of a conical hill that we met earlier. Drop a speck of dust into the water, and it acts as a nucleus for the formation of ice crystals. The result is a sudden, almost explosive freezing of the whole water.

When water freezes, it gives out heat. The molecules in water are all buzzing around and colliding, but in ice, they are confined to vibrate in their lattice positions. The kinetic energy of these molecules has to go somewhere. It is given out as latent heat to the surroundings. (The reverse happens when ice melts. You have to supply a lot of heat to turn ice into water.)

Exactly the same thing happened with the universe. As it cooled, the weak nuclear force and the electrical force should have separated out from each other and broken their symmetry, but for a time, the universe went on cooling beyond this point, with the symmetry still intact. It was now in a highly unstable state, for if it could give away its excess energy, it could then exist in a less symmetrical state. Suddenly, the symmetry broke, and at that moment, an enormous amount of "latent heat" was released. Like rapidly freezing water, the already expanding universe suddenly expanded in a dramatically new way called "inflation." Unlike the normal expansion of the universe that slowed down with time, this rate of explosive expansion increased with time. This event lasted only an incredibly tiny fraction of a second, but during that period, the universe increased in size by a factor of 10^{30}!

From that point, everything was downhill, energetically speaking. The universe passed though the era of hadrons, then into the leptons, and, as we saw in the start of this chapter, it finally cooled down into our extended universe, in which the first galaxies and stars formed.

This story is widely accepted within the physics community, although it has a number of problems and unresolved details beyond the scope of this book. But two of them are worth mentioning. One is the anthropic principle, and the other is the ultimate question of all: How did the big bang itself originate?

The Anthropic Principle

The universe contains a number of constants of nature, such as the speed of light, the charge on the electron, the ratio of the weak nuclear force to the electromagnetic force, the cosmological constant (a feature of general relativity), and so on. Physicists believe that an ideal theory would explain the size of all these numbers and leave nothing arbitrary or up to chance.

At present, such a grand, unified theory does not exist. Physicists just have to accept the experimental values nature gives them. There are also elements of the big bang theory that require physicists to adjust some parameters in order to have the theory come out right. Physicists don't like doing things this way because it means that their theories are arbitrary. There is, for example, no compelling reason that the ratio of the electromagnetic and weak nuclear forces should be one particular number rather than any other.

And so, scientists play around with these numbers, testing out their theories to see what would happen if one of these constants is changed by the tiniest amount. It turns out that the world in which we live could not exist unless all the constants of nature have almost precisely their present values. It is almost as if, out of all the possible values of these constants, only those that we now measure can give rise to a universe in which there are stars, planets, and intelligent life. For example, if the charge on the electron happened to be a little different from the known experimental value, then the stars could never burn hydrogen and the universe would be cold and dark.

This apparent adjustment of the constants of nature is known as the anthropic principle, and its meaning has given rise to considerable speculation. It has been stated in many different ways (again, Dionysus and Apollo are hiding around the corner). One is that the universe was specifically designed for us to live in. If the constants had been any different, then planets, human beings, and consciousness would not have been possible. Another version holds that an infinity of universes exist, all with different values for the constants of nature; only in our own universe do the constants, by chance, have exactly the correct value to produce life and beings capable of wondering about these things. Yet another approach pictures an infinite universe containing many small, local fluctuations. According to this notion, our own "universe" turns out to be just a small fluctuation in this giant cosmos—one that just happens to have the correct adjustment of constants necessary to produce life.

At the moment, the anthropic principle is a mystery. Maybe a deeper theory will account for all these constants perfectly. Maybe the constants are not fixed at all but are merely an averaging out of something deeper. It could be that our present universe is the result of self-organization at some other level. Or maybe we live in an evolutionary universe and the constants have changed with time. There is even a bizarre science fiction–like suggestion from the cosmologist Fred Hoyle.

Physicists know that in quantum theory, the act of making an observation changes a system. Some physicists have argued that, in what are called "delayed choice experiments," this disturbance can be made retroactively. This means that by acting now, one can change a quantum state in the past. Hoyle's notion is that highly intelligent beings of the future develop the technology needed to make retroactive, delayed choice observations within the first instants of the big bang. In this way, they can disturb and tweak conditions during the big bang to produce just the sort of universe they want.

In other words, these beings of the future act back in time to create a universe in which life can evolve and ultimately produce the same intelligent beings that are now making the observations! Another age may have invoked God as the agent of perfect creation, but we appear to require quantum engineers. The results are much the same: We attempt to resolve mysteries by projecting onto them the products of our all too human imagination.

The Big Bang

The other great mystery is the big bang itself. Even at the level of the unification of gravity and the world of superstrings, physicists are out of their depth in this regard. The ultimate origin of things is an even wider ground for speculation. I call it speculation because the world of physics is still waiting for a truly fundamental theory that will unify relativity and quantum theory. By this, I don't mean a theory that tries to force them together—this has already been tried without much success. I mean a much deeper theory out of which relativity and quantum theory emerge as limiting cases. Just as in relativity theory Newton's world emerges as a limit for speeds slower than light and small gravitational fields, so too will a much deeper theory contain space, time, and matter, as well as relativity and quantum theory, as its limits.

So what are we left with? Maybe the universe began by pure chance. Maybe it was no more than a quantum fluctuation: an expression of the potential of the quantum vacuum that enters the manifest world of space and time. The problem with this notion is that quantum theory is essentially an incomplete theory. It is not possible to define a wave function on its own; it must instead be defined only with respect to a particular observation or experimental context. Quantum theory explains the classical world as a limit of the large, but it needs this same classical world in order to define its wave functions and explain their collapse.

What sense does it make to have a wave function for the universe when there is no observer? How can such a wave function collapse and achieve the actuality of a given universe out of an infinite number of potentials? People have tried to patch answers together; they've attempted stopgap solutions. Some of them are ingenious—Stephen Hawking's book B*rief History of Time* explains these sorts of approaches. Yet they're not really founded on anything particularly deep. They are really makeshift ways out of a critical situation in theoretical physics, not profound answers in themselves.

What else is there? The physicist John Wheeler speculated that neither space nor time nor matter but rather logic is "the nuts and bolts of the universe." But what sort of a logic would it be? Not the traditional logic we inherited from the Greeks, one in which every proposition is either true or false. An example of such logic would be the statement "an electron is either a wave or a particle." Quantum theory deals in complementarity ("an electron is both a wave and a particle"), and the answers one gets depend upon how the question is asked. Such a logic would have to be context dependent. A statement will be true or false or ambiguous depending upon the context in which it is placed. Expressed in abstract symbols, such a logic may provide the basis for new sorts of theories.

Another physicist, David Bohm, argued that despite the revolutions of relativity and quantum theory, physics still clings to the old order first proposed by Descartes. This is the order of space described by coordinates and infinitely divisible into points. For Bohm, quantum theory points to a radical new order in which things can enfold into each other. This is an order in which A can be contained within B at the same time that B is contained within A. Bohm called this the "implicate order," and with his colleague Basil Hiley, he worked on algebras that would express this different way of seeing the world. (These algebras were first developed in the nineteenth century by the mathematician Grassman as a way of describing the movement of thought in algebraic terms.)

At the time of this writing, only a handful of imaginative and creative physicists are working on genuine alternatives to current theories. As yet, none of their new ideas have really paid off. And so, the mystery remains. We don't really know what happened at the big bang because we don't yet even possess the language in which to properly express such ideas. As Hamlet put it with his dying breath, "The rest is silence," which seems a good way to move on to the next chapter.

chapter 4

Silence and the Void

In the previous chapter, we learned of the incredible activity that took place during the birth of the universe, with entire eras of cosmic evolution occurring in less time than it takes for a camera shutter to click. Yet the essence of creativity need not always lie in incredibly fast processes and change. It can also be found in silence and absence.

It is easy to fall into the trap of believing that if we are truly creative, then we must always be involved in some sort of activity: that we must be planning, controlling, working, and producing. We tend to think that we're being most productive when making something or putting some sort of plan into action. However, we can also be incredibly creative when we are silent, doing nothing, and suspending all action.

To illustrate the point, I'd like to begin this chapter with a personal anecdote. Five or six years ago, I felt that I could no longer come up with any new ideas. The work I do and the various avenues I explore are all supported by royalties from the books I write. My bank overdraft told me that it was time to work on a new book, but I simply could not think of anything to write about.

Around that time, my wife and I decided to sell our house in Canada and move, for a time, to Europe. The result was that we took a three-month break in Italy, staying in a small, medieval, hilltop village. We arrived in the heat of September, and for the next three months, I spent most of my time sitting in a chair outside the house and just looking over the hills. Sometimes, I'd get up and go for a walk, but most of the time, I'd just sit and look.

Later, when I met old friends, they'd ask me, "What did you do in Italy?" "Nothing," I'd reply. "Yes, sure," was their answer, "but what did you *really* do? Did you read a lot? Did you do any writing? Were you painting? What on earth did you do for three months?"

No one seemed to believe that I did nothing. Of course, at times, it was hard to resist the impulse to begin to write things down on paper or to go into Siena and buy some books to read. But, apart from a few lapses, I mainly spent those three months just sitting and looking. I suppose I must have been a little like a woman in her last weeks of pregnancy. I really had no idea why I was doing nothing, but something deep inside me, something I really trusted, told me that it was the right thing to do at that time.

When the break was over, we moved to London for a year, and suddenly, I found that all sorts of ideas were beginning to surface, half formed. Now, five years later, I'm writing this present book in the same village where I spent that time of withdrawal. I'm writing every day yet keep having to take time out to pursue other ideas and projects. I've got so many leads to follow up and so many ideas to explore that I simply don't have time for all of them. Everything that is happening now grew out of that period of silence and inactivity.

By this, I don't necessarily mean that the ideas themselves were present as seeds during that fallow period and that they had simply needed some space in which to germinate. No, I joked to myself at the time that I was making compost. I was a sitting, walking compost heap. During that period, I was making the fertile soil that would nurture the seeds that were to come. Then, in the following years, as the seeds fell on the soil, those gifts from the universe had somewhere to grow and be filled with energy.

The power of doing nothing is the power of silence. It is the untouched marble that, to the sculptor, represents pure potential. It is the moment of silence in a piece of music that contains, enfolded, the whole work. It is that void seen by the mystics and, as we shall see toward the end of this chapter, the state of total absence in quantum physics that is paradoxically the most full.

Void in Art and Music

The total absence—the negation that, at the same time, is a plenum and an absolute fullness—is found in the writings of mystics. It is also the vision expressed by artists and musicians alike. All that exists outside the void and in the domain of time is, to some extent, conditioned. For the Buddhist and physicist alike, to be tied to the wheel of time is to be caught up in the eternal web of cause and effect. The void, the negation of all, lies beyond this causality. It is unconditioned, pure potential.

That which exists in the manifest world and within the domain of time can be defined and named. Within the mind, a boundary can be drawn around it so that it can be gathered together as a concept or idea. But the void lies beyond all names and all definitions.

From ancient China comes the notion of Tao, or "The Way." The Tao is that which has no name, that which lies beyond naming:

> The Tao that can be spoken is not the true Tao;
> The Name that can be named is not the constant Name.
> The Tao without name—that which is non-Being—was the beginning of Heaven and Earth.
> That with can be named—that which is Being—is the mother of the ten thousand things (author's translation of a passage from *Tao Te Ching*, or *Book of Changes*).

To the thirteenth-century theologian Meister Eckhart, that which could be named and spoken of was not God but only our human image of God. For many mystics, God was seen as a radiance and divine light. But for an uncompromising intellect like Eckhart, the divine had to be approached through the *via negativa*. By saying what God is not, we gradually strip away all concepts, all definitions, all mental images, and all attempts to name, contain, and reduce. Finally, when everything has vanished, we are left with the void. We are left with nothingness, with that which lies beyond all definition. In a Zenlike paradox, the mind is left with the ungraspable, with the unthinkable, with that which is neither a noun nor a verb nor any other part of speech. It is a state without image or reference, a state of mind beyond thought and beyond thinking. And if an image is required, then it is that of absolute blackness, an absolute absence that is, at the same time, a plenitude of radiant light.

Quantum Vacuum

In a particularly striking way, this notion of the void is also present in modern physics in the guise of what is known as the quantum vacuum. Let us begin this investigation of the power of silence by asking what happens if we take all the matter out of a region of space. Is it then empty? No, it is rippled with energy, with rays of light and waves of gravity. But suppose even these are removed. Suppose that all matter and all energy is drained from a zone of space. What then?

Quantum theory acknowledges such a state of nonbeing. It is called the "vacuum state" or "ground state." It is that state out of which all matter and energy have been abstracted. It contains nothing. It is pure, absolute emptiness. This is the quantum theory metaphor for that void sought by the mystics.

Paradoxically, this void, this total absence, is at one and the same time totally full. The nature of the quantum void is another of those surprises with which the quantum theory confounds our common sense. At another level, however, it makes a metaphoric connection to more ancient wisdom.

When your car battery has gone dead, you can get no more energy out of it. When a fire goes out, it will supply no more heat. When a clock runs down, it no longer ticks. Left to themselves, things in the universe run down until they have lost all their energy. In quantum terms, this is the end point known as the vacuum state. The vacuum state has lost all its energy and can achieve nothing more.

But here is where the paradox comes in. The central premise of quantum theory is uncertainty. At the quantum level, some pairs of properties cannot be known for certain at exactly the same time. For example, if we pin down the exact position of an electron, then we become uncertain about its speed. In one sense, the act of trying to measure the speed of the electron actually disturbs its position in an uncontrollable way. But the paradox goes even deeper, for quantum theory asserts that the very concept of an electron with a definite speed and position can no longer be maintained. Uncertainty goes much deeper than the fact that our experiments act to disturb the universe. It is a fundamental property of all quantum processes.

Another of these uncertainties involves energy and time. The more we attempt to pin down the energy of a quantum system to within a precise instant of time, the more this energy becomes uncertain. In our everyday world, none of this has any practical significance, but within the lifetime of the shortest-lived elementary particles, this uncertainty can have strange effects.[1]

How does this apply to the vacuum state? Saying that the vacuum state has no energy turns out to be a very precise definition because it tells us that the energy is exactly zero. This does not matter if we're only concerned with the zero energy of the vacuum state to within a second. But what about the energy within, say, the lifetime of an elementary particle? When

---

[1]For the scientifically minded, this discussion is somewhat simplified. Strictly speaking, uncertainty applies to what are termed "noncommuting observables" (for example, position and momentum). In conventional quantum theory, time is not an observable but a parameter in the equations. So though it is true that uncertainty exists between energy and time, its formal origins within the theory are not as clear-cut as those between, say, position and momentum. Needless to say, a number of physicists have tried to clarify this point.

we try to identify the precise energy in such a tiny time interval, we find that it becomes noticeably uncertain.

Another way of saying this is that the vacuum state is constantly fluctuating. Sometimes, it is exposing a little energy to the outside world; sometimes, it is taking it back again. Over a lifetime, a minute, or a fraction of a second, this energy averages out to zero. The vacuum state is truly empty, truly a void. But what happens when we reach down into incredibly tiny intervals of time? Then, we discover that energy is really fluctuating in an unpredictable way.

The theory of relativity teaches us that energy and matter are interconvertible. If you have an amount of energy within a region, then this energy can always turn into matter. Having energy within a region of space means that an elementary particle can suddenly burst into existence.

The vacuum state is the void. It is pure silence. But it is also a bubbling sea in which elementary particles are constantly dancing in and out of existence. A good analogy is the money you get out of the automatic cash machine. Suppose your bank account is empty. You have zero money, but because the bank's computer has developed an electronic glitch, the machine allows you to borrow twenty dollars. The computer program isn't going to discover the deficit until the end of the day. So now you can go to the corner store and buy bread, a can of soup, and some coffee, provided that you're able to borrow the money from a friend and pay it back before the bank closes. As far as the bank's accountants are concerned, your account remains at zero and nothing has been borrowed or paid back.

Now suppose the machine gives you two hundred dollars! You're free to go out and spend all that money. Provided you pay it back within the hour, the bank's computer won't discover that anything is wrong. Then one amazing day, it issues you ten thousand dollars. You can run out and buy a new car— provided that you pay back the money within sixty seconds! The faster you pay back the money, the more you are given. But the longer you keep the money, the less you can sneak out of the machine.

Exactly the same thing happens when an elementary particle moves through the quantum void. In place of you and your bank card, read an electron, and for money, read energy. As the electron moves, it's constantly negotiating with the bank machine of the void. The more energy it borrows, the faster it must pay that energy back. Borrowing large amounts of energy for a very short time means that this energy can be turned into matter, and thus, the electron is constantly transforming itself into other particles and back again. Averaged over a long period, the electron remains an electron but moves into tiny fractal dimensions of time, and it is in a constant state of transformation.

Now imagine a positively charged electron and a negatively charged electron meeting in a collision. They disappear into a burst of energy. But they can use that energy to create themselves again, so those two electrons then fly away. But what if, within the tiny fraction of a second of their collision, they "borrow" a little more energy from the vacuum? They now have enough energy between them to transform themselves into a positively charged proton and a negatively charged proton or some other pair of particles. In this way, they are constantly changing their identities—now particle of light, now electrons, and now protons. Throughout their lives, their identities are never fixed. All elementary particles have multiple personalities.

The shock comes when we realize just how much energy can be borrowed. It's not just a matter of a bank machine attached to the local branch of our bank but a bank machine wired directly to Fort Knox! Provided that energy is borrowed and paid back fast enough, the amount of energy available is unlimited. It is at this point that we begin to realize that our commonsense picture of reality has been turned on its head. Rather than reality consisting of molecules, atoms, and elementary particles connected to some insubstantial vacuum state, it is the vacuum state that becomes the primary reality. The world of electrons, atoms, planets, and stars is no more than a tiny bubble upon a vast sea of energy possibilities.

Our physical world is a little like the clouds in the sky. Seen from the ground, clouds look like substantial objects moving above our heads. But as you enter the clouds while flying in an airplane, you discover no hard-and-fast boundary around them. They simply dissolve into a fine mist. The same is true with our world. It appears substantial, yet, at base, it is just an endless series of fluctuations, and with respect to the vacuum state, its existence is totally insubstantial.

Suppose all the matter in the visible universe is annihilated and converted into energy. It turns out that the energy within one cubic centimeter of the vacuum state would vastly exceed the energy content of the entire universe. And what's more, if all the matter in the visible universe were to be converted into energy in an enormous cosmic explosion, this energy would still be insignificant when compared to a cubic centimeter of the vacuum.

So this void, this nothingness, this cosmic silence, is pure potential. It is an action suspended, an action waiting to be called into manifest existence. The physicist John Wheeler compared the quantum principle to the Merlin Principle. Merlin was the wizard who guided King Arthur. Sometimes, he would appear to the king as a young boy; at others, he would be a wise old man or an animal. Just like the electron, Merlin was constantly changing his identity.

The electron is able to change its identity because it is in contact with the void. If it were isolated and unable to touch the ground of its being, then it would remain simply an electron. It is almost as if the electron were exploring the fractal dimensions of time. Over long time intervals, the electron remains itself, but as we focus our temporal microscope on finer and finer intervals of time, we discover that the electron is behaving in ever more bizarre ways, transforming itself into all manner of unusual elementary particles. To borrow an image from the poet Gerard Manley Hopkins, by virtue of its connection to the void, the electron possesses an *inscape*. It has a rich inner nature that can only be understood by looking at it at increasingly smaller time intervals. In another sense, the electron becomes a vehicle through which the creative energy of the void is flowing. The electron is no longer an object in space and time but a process though which the void manifests itself within the domain of space and time

Connecting to the Void

This same sense of the infinite creative possibilities inherent in contact with the void is also found in art and mysticism. Artists have pursued the concept of the void and the unnamable. During the eighteenth and nineteenth centuries, the idea was related to notions of the sublime in poetry and painting. In place of the earlier fashion for the merely picturesque came an interest in vast spaces and sweeping vistas—around this time, the English discovered that mountains were for walking in and enjoying, rather than troublesome barriers to travel. The sublime was the province of mountain peaks, oceans, cataracts, dark caverns, and the infinite depths of space.

Instead of appealing to our conventional sense of beauty or to some ideal of balance and calmness, the sublime evokes feelings of awe and even terror. In its presence, human beings are dwarfed to the point at which their concerns seem insignificant and petty. The sublime overwhelms the mind to such an extent that the more subtle and delicate aesthetic elements in art are in danger of being neglected.

The sublime makes explicit the enormous power of nature. In early times, people lived more directly in nature and discovered ways of forming relationships with its powers and energy. For this reason, they had no need to invent such a poetic notion; the sublime is, therefore, more a conceit of the city dweller, someone who comes into contact with nature from the comfort of a carriage while taking the grand tour!

In turn, the sublime found its way into romanticism, with its spirit of revolt against authority and the heroism of the individual who shakes a fist against the gods. Romanticism was certainly a glorious period in art and literature, yet our own age has seen the dangers that follow the abdication of reason in favor of impulse and the forces of the unconscious. We have become more guarded, and so we view the void in a more cautious manner.

In the following sections, we shall see the different way in which twentieth-century artists and composers approached the void. Let us begin with Anish Kapoor, a contemporary artist who works with the void in a way that, though rooted in earlier sensibilities, is nonetheless truly modern. Many of Kapoor's works contain fathomless depths. To understand their nature, it is first necessarily to distinguish the void from a hole.

It was the British sculptor Henry Moore who realized the significance of the hole in sculpture (along with his contemporary Barbara Hepworth). There have always been holes made in sculptures, but for Moore, the hole became a new means of linking spaces together and allowing the eye, as well as the felt sense within the viewer's body, to journey in a radically new way.

The hole interpenetrates space and in doing so links surfaces together. Moore's work is almost akin to a visual paradox. As you walk around the piece, the eye and body travel round the surface of the stone, forming a felt sense of the image. But at the same time, one can transcend space and time by entering the hole and linking immediately to the other, as yet unseen space behind. (Moore himself has written, "To understand real three dimensions is to train your mind to know when you see one view what it is like on the other side, to envelop it inside your head, as it were. We learn distances originally by walking them. We understand space by understanding form" [*With Henry Moore: The Artist at Work*, 1978].)

For Moore, the hole links two equivalent spaces to create a sense of a space that lies beyond them both. The voids in Kapoor's sculpture do not work in the same way. They do not link the front to the back of the sculpture in order to bring us to some new space beyond. Rather, Kapoor's cavities are entry points into a space that is radically different, a space that is of a quality other than dimension, mass, and surface.

His *Adam* consists of a large block of naturally mud-colored granite, and it reminds us that the first man was made out of mud and given life by the breath of the creator. The monolith has a deep hole carved at the height of the viewer's head, and the inside of the hole is coated with deep-black pigment. This use of raw pigment, characteristic of much of Kapoor's art, evolved from earlier pieces in which powdered pigment in primary colors was tossed over his sculptures, evocative of the Hindu festival of Holi.

In *Adam* and Kapoor's later works, the pigment is used to a dramatically different effect. The artist applies the black pigment using a solvent. After each layer dries, he applies a successive layer, using less and less solvent carrier until, after some twenty applications, the final coats are nothing more than pure pigment without solvent. The result is dense and light absorbent. All visual clues vanish, and one who contemplates the work enters a void of pure blackness, an aperture with no end, dimension, or surface.

The void occurs again and again in Kapoor's work. Sometimes, it is in the guise of black marble that has been heavily polished to create a concave inner surface that is both absorbent and reflective. Other works consist of large hemispheres, bigger than a person, whose centers are coated with blue pigment; one is left in the face of an endless blue, a blue that sucks in light (and, incidentally, sound, as well).

Kapoor's work is particularly striking and has evoked the term *beauty* among critics and art historians—a word rarely used in connection with contemporary art. However, one's immediate response is not aesthetic in an analytic sense but deeply visceral. In some ways, his pieces are evocative of that earlier notion of the sublime. When we stand before a piece of sculpture, we normally place ourselves in relationship to it, to its scale and mass, to the way it occupies space and causes us to create our own spatial relationships, to the ways our own bodies occupy space in the face of the work.

Sometimes, the way we create this space in which "to be before the work" requires us to move around and understand the work as a form of process. Other pieces invite us to discover a place to stand, from which we can confront or relate. Kapoor does not allow us such a convenience. In front of the void, all visual clues are missing.

As we approach his work, we may have a sense of fear, of losing orientation. One first comes to Kapoor's works with a tightening of the stomach, as the body attempts to maintain its equilibrium. One feels the type of vertigo experienced at the edge of a cliff, where there is no sense of firmness on which to make the next step.

I once observed a young couple standing before one of Kapoor's blue hemispheres. Their engagement was intense, and I noticed the young man glance at his girlfriend, almost as if to assure himself that she was still standing beside him, still within the gallery space, and not sucked into the void.

In facing Kapoor's voids, one is facing the void within oneself, the inner darkness, the absence, the pure potential. It is tempting to ask where the work exists. Is it in physical pigment and stone? Does it only exist within the mind of the viewer? Or is the work that space between, that process that arises within the act of seeing?

All too often, our lives are filled with activity—lists of thing we must do on the way home from work, plans that must be implemented, insults that have to be paid back, desires that remain unfulfilled. In this way, a large part of our lives may remain unlived; we can find ourselves absent from the present because we are half living in the past and half anticipating the future. And so, the mind churns on in its ceaseless and often pointless activity.

Kapoor's work abruptly cuts the ground from under our feet. Its impact is so direct, so visceral, and so unmediated that we are arrested by it, and for as long as we stand before the piece, we remain in contact with a deeper aspect of ourselves. We tend to think of inner wisdom and insight in terms of enlightenment or of the light of the intellect. But the void contains within it the resolution and mystical marriage of all dualities and contradictions. Kapoor's works therefore explore the *via negativa* in terms of endarkenment.

By contrast, one can also approach wisdom though light, and another series of Kapoor's works deal in dazzling whiteness. The interiors of his boxes are bathed in such diffuse white light that it is impossible to detect their interior dimensions. Entrances into blocks of marble are excavated so deeply that only a fine membrane is left at the back, and the translucency of the stone illuminates its interior. Another Kapoor series features highly polished steel, so that the entire surface of a piece dissolves into such reflections that inner and outer becomes confused and united.

A related exploration of the void as pure light comes from music. In *Ikon of Light*, the contemporary composer John Tavener expresses the duality of absence and fullness in terms of a plenitude of light. The work is a setting of verses by the mystic Saint Simeon (the New Theologian of the Orthodox Church) that contemplate the nature of "uncreated light." The work begins with the word *Phos* (light) and ends on *epiphania* (a shining forth). In one sense, the composition involves a form of synesthesia, the expression of light in terms of sound: light as both a physical illumination and a spiritual shining forth. Light builds in intensity though the simple musical means of repetition and strict counterpoint. At the center of the work, the vision intensifies and the music shimmers. As instruments and choir augment the layers of light, the element of silence, between the verses, becomes increasingly significant. Within the greatest intensity of light appears absence, absence of all sensation because the vehicle of vision has nowhere else to go. Within the multiple layers of sound is pure silence, pure potential. The void becomes a state of holding, the suspension of all movement.

The overwhelming vision of light recalls Dante's vision of God as a point beyond spatial dimension in Canto XXVIII of the *Paridiso*.

> I saw a Point that radiated light
> Of such intensity that the eye it strikes
> Must close or ever after lose its sight
> —Dante Alighieri, *The Divine Comedy*, trans. John Ciardi (1987)

Tavener himself has attached the following quotation, from the Russian Saint Seraphim of Sarov, to the score: "I can't look at you, Father, because the light flashing from your eyes and face is brighter than the sun and I'm dazzled."

So, as with the vacuum in quantum theory, the void for Tavener and Kapoor becomes both a plenum and an absence. A few decades earlier, it had also been explored by the French artist Yves Klein. As a youth, he had specifically claimed the void for himself. Once, while sitting at the seashore, he and his friends divided up creation. One of them, Claude Pascal, laid claim to words, while Armand Ferdandez plumped for the earth. Klein, however, chose *le vide* (the void, vacuum, empty space, sky): "I have written my name on the far side of the sky!" he said (quoted in Hans Weitemeier, *Yves Klein, 1928–1962*, 1994).

Klein expressed his experience of the void through large, monochrome paintings. In particular, he spent some time developing an ultramarine pigment and solvent medium that came to be known as International Klein Blue (IKB). His monochromes were not hung flush to the gallery wall but suspended almost a foot away, so that they appeared to hover in a zone beyond mass, dimension, and location.

Another of his ventures, which became specifically known as "Le Vide," consisted of a totally empty gallery room. Prior to the exhibition, Klein spent two full days in a meditative state of total emptiness, painting the gallery walls white and using the same pure pigmentation approach he employed for his monochromes. He then said that he had filled the gallery with the essence of the color blue, not its physical manifestation as pigment or colored light but as that which transcends all "dimension of color" (Weitemeier, *Yves Klein, 1928–1962*, 1994).

The records of that exhibition indicate that it was by no means a gimmick. Many of the leaders of French cultural life were moved by the show, and the event, along with Klein's other work, has had an important influence on the subsequent history of art. Albert Camus wrote in the visitor's book, "Avec le vide les pelins pouvoirs" ("with the void, full powers"), and the philosopher and scientist Gaston Bachlard presented the artist with a copy of his poetic meditations on the nature of sky and void, *L'Air et les Songes*. For Klein, this exhibition heralded the era of pure spirit in art—the pneumatic epoch, in which art would transcend aesthetics and move into pure consciousness and "the one thing that does not belong to us: our LIFE." This was the same Klein who had said of his work, "My paintings are the ashes of my art," and who shortly before his death wrote, "Now I want to go beyond art—beyond sensibility—beyond life. I want to enter the void" (quoted in Weitemeier, *Yves Klein, 1928–1962*, 1994).

The sculptor Antony Gormley explores the void from the perspective of an experiential, inner space. As a child, Gormley was sent to his bedroom to rest each afternoon. At first, the experience was claustrophobic, but as time went on, he realized he was able to enter an internal space that began to expand without limits.

As an adult, within the exercise of his art as well as through meditative practice, Gormley has been exploring this inner space. He describes it as "the inner space of the body; a space that is rarely visited. It is a space that lies beyond dimension, beyond up and down, beyond good and evil, yet contains them all" (Gormley, in conversation with the author).

In the creation of many of his works, Gormley is first encased in plaster while in a meditative state—the process of the work is somewhat like a death and subsequent resurrection. The resulting plaster cast forms the basis for a cast made of iron or some other material. Or it may serve as the starting point for further work, as with his massive *Angel of the North*—the largest sculpture in Britain—in which the arms have extended to suggest wings or that inner experience one has of being able to stretch out and touch the very edges of space.

The origin of Gormley's pieces lies within a private inner space. Yet the final works occupy external, physical, communal space, and they do so in particular ways. Some are poised in perilous positions of balance. Some confront each other so as to define and delimit the space around them. In one exhibition, the gallery was empty and the sculptures were located outside, placed next to trees or standing on the roadside. The work was designed to be viewed from within the gallery space but though its windows.

Gormley approaches the void in a different way from Kapoor and Klein. Standing before a humanoid figure, we subtly and largely unconsciously place ourselves in relationship to it by making subtle body adjustments and reorientations. In doing so, we become aware of the way we also occupy space and how we stand in relationship to the piece and the walls of the gallery. Of course, we are always doing this, every time we enter a room full of people. But we are not normally aware of how we occupy space. Or, to put it more accurately, we do not give attention to the way our inner, subjective space merges horizons with the external, communal space of a room or a crowd.

But because Gormley's pieces, as works of sculpture, are other than human, they alert us to what is going on and cause us to become more sensitive to our inner space and to its relationship to the external. If Gormley's work could be said to have an aim or goal, it is to bring the viewer to an awareness of the space within each one of us—that "space rarely visited"— through its very physical presence (mass, material, surface, orientation, form, and occupation of space).

To return to Klein for a moment, his radical approach, with its rejection of many of the tenets of Western art, may, in part, have been the result of a journey he made to Japan to study judo and its associated mental disciplines. A similar Japanese influence, specifically Zen, is present in the work of the composer John Cage. Cage was aware that Western music, particularly since the romantic period, had focused on heroic, biographical pieces. Nineteenth-century musical structure, with its tonal tensions and resolutions, its contrasting symphonic movements, and its climaxes and codas, is a mirror of this attitude.

Cage's music points in a quite different direction. It explores the dual nature of the void (absence and plenum) in terms of music (silence and noise). One of his best-known pieces is *4' 33"*, a work in which the pianist sits at a grand piano, opens the lid of the keyboard, and then sits in silence for exactly four minutes and thirty-three seconds. The pianist then closes the lid of the keyboard. The piece is far from being a stunt or joke.

In one sense, it exposes the power of silence; in another, it demonstrates that silence is never truly present. Like the quantum vacuum that is never truly empty, silence is never totally quiet. All around us are sounds that themselves could be the raw material of music.

Let us begin with absence within the sonic void. Although others have explored musical sound, Cage was more interested in the intervals between notes, the silences out of which music emerges. In search of absolute silence, he entered an anechoic chamber—a room built for acoustic research that absorbs all sounds. After a short time, Cage realized he could hear two distinct sounds. Afterward, the scientists told him that one was the pulsation of his blood and the other, higher pitched was the firing of his nervous system. So within the heart of absence is presence—the plenum of noise.

Aware of this ambient sound, Westerners have traditionally attempted to bracket music during the act of listening. Music is taken as an expression of art and different in quality from all that is nonmusic: the sounds of the environment, accidental sounds made by the physical act of playing musical instruments, and so on. In Japanese music, by contrast, the sounds of the flutist's breathing, the fingers tapping on the instrument's key, and so on are all appreciated as part of the music. Cage approached music in a similar way, with no attempt to give more emphasis to the "musical" than to all other sonic events occurring in the concert hall.

Yet, in the last analysis, even this was not sufficient, for the ego of the composer is always present, making aesthetic judgments about the aesthetic appeal of sound. This dilemma and his own study of the *I Ching* led Cage to incorporate chance processes into the composition of his music, such as throwing dice or making use of tiny imperfections in a blank musical score to insert notes.

For Cage, the void is pure potential that lies beyond object, concept, and ego control. It is that which cannot be encompassed, isolated, and bounded in thought. Thus, he invites us to "see" the world around us with our ears. This approach is echoed by two contemporary artists, Madelon Hooykas and

Elsa Stansfield, who have constructed a series of "observatories" to bring the visitor into direct contact with the world in an unmediated form.

Their sculptures or installations are frames placed around an experience. Just as with Cage, who isolated four minutes and thirty-three seconds of sound, or Klein, selling a specific zone of empty space, the two artists bracket our experience of the world. Their observatories are markers to remind us of what we are now experiencing. They are like those bells periodically rung in Eastern monasteries to remind the monks that they are meditating.

One of their observatories is constructed on sand dunes beside the sea in Holland. It has the appearance of the parabolic mirror of a radio telescope. Visitors climb to the observatory and sit within the mirror. This brings their heads into such a position that they can focus on all the sounds of the sea. What looks like a radio telescope is, in fact, a telescope for sound, with the noise of the wind and waves, the cries of gulls, and the shouts of children going right into their ears. Those who experience the sculpture have, of course, always been bathed in sound, but when those sounds are isolated into a zone of focus, they can be attended to with greater awareness. Thanks to the observatories, we realize that we are immersed within the void of nature, the plenum, the silence, the activity of all that is.

In the previous chapter, I wrote of the wonder I experienced on first learning that we are connected directly to the distant stars, since the proper time taken by a photon of light to travel from a star to our eyes is zero. But there is something even more outstanding in all this. When we look up at the night sky, we realize that light is reaching us from a great area of sky. This light is, in fact, coming from all over that region of the universe—the angle subtended at our eyes—and we are receiving light or information from tens, hundreds, and millions of light-years of space. We receive information about billions of stars and galaxies and about the way rays of light have been bent and twisted by gravity during the journey to our eyes. In the blink of an eye, we receive an enormous quantity of information from the heart of the cosmos, from the depths of space, from "le vide" claimed by Yves Klein.

In turn, that information becomes transformed into chemical processes at the retina and then into electrical signals that flow up the optic nerve until they unfold over the visual cortex to produce our "seeing" of the cosmos. These same processes are repeated when we hear the sea in Hooykas and Stansfield's observatory, listen to the music of Cage or Travner, or contemplate a Klein or a Kapoor.

A composer such as Jonathan Harvey approaches the void from within the plenitude of a note or a single musical event. Using electronic processing, it is possible to sustain an instrumental (or electronically produced) note indefinitely. No single note or sonic event is pure. A note contains a variety of overtones, in addition to transients and resonances produced by the way the note is musically attacked, is built, and then dies. Add to this the resonances with the physical material of the instrument itself, as well as those from the room in which it is played, and a single E on a violin carries within it an enormous inscape of sound.

In Harvey's piece *Bhakti,* sounds made by a musical ensemble are manipulated electronically to give emphasis to certain aspects of a note's inner life and the music's interiority. Unlike conventional Western music, the piece has no goal or linear development. It is a meditation upon pure sound and pure potential. Sections from the *Rig-Veda* are associated with each section of the work, beginning with a quotation that perfectly illustrates the nature of the void: "There was neither non-existence nor existence; there was neither the realm of space nor the sky which is beyond. What stirred? Where? In whose protection?"

This plenitude within the inscape of sound, this possibility for a total fullness without movement, is present in much Eastern musical tradition, as, for example, in Tibetan overtone chanting. It is a renunciation of the Western desire for progress, movement, change, and ceaseless activity, for a constant effort to get away and escape the moment, the now. It is music that remains with the now. It is of the moment, immediate experience, with what is given. It is an act of watchful listening without any attempt to cling to sound within transitions.

A composer such as Cage knows that silence is enormously powerful. That is why so many people find it difficult to tolerate silence and try to fill it with talk or the acoustic wallpaper of radio and television. (I once knew someone who, on entering his house, always switched on the radio on his way across his living room to turn on the television.) The mark of a deep friendship is that ability to sit in silence together for long periods without the need for speech.

The first sessions of psychotherapy may be filled with the patient talking, struggling to get her story told, and trying to satisfy what she fantasizes the therapist wants to hear. In those early days, periods of silence are always being filled with words. Finally, the time may come when the patient falls silent. If the therapist is strong, then this silence can be contained and become an alchemical vessel in which a new kind of work can take place.

Bach's *Passion According to St. Matthew* is the most moving of his works. A shimmer of sound from the continuo surrounds Christ's words. It is only with the final words said on the cross—"It is finished"—that the accompaniment ceases, allowing the words to be uttered into a void.

Benjamin Britten used a similar device in his opera *Peter Grimes*. The fisherman Grimes is consumed with dreams and ambitions. He will "fish the sea" to prove his superiority to the citizens of The Borough. But now, he has reached the end of his tether. Two of his apprentices have died in suspicious circumstances, and a third has fallen (or been pushed) down a cliff face. As the citizens issue a hue and cry, Grimes, half mad, has nowhere to turn; only his friends Ellen Orford and Captain Bulstrode remain loyal. Bulstrode realizes that Grimes can only find peace in death. In a highly dramatic moment, he tells his friend, "Sail till you lose sight of the Moot Hall. Then sink the boat." For the first time in the opera, the words are spoken, not sung.

But the most moving of all silences in music was never planned. It occurs during the final contrapuntus of Bach's *Art of Fugue.* A second theme has entered, based on the musical correspondence of the letters B-A-C-H. Then, as the theme enters afresh, the music abruptly ends. Bach died before he could complete his great work. This instant of suspension is one of the most creative opportunities in music, maybe in all of art. If we are able, we can continue listening in absolute silence, letting the unsounded music play itself out in our minds and bodies until it fades or until the fugue ends as Bach intended.

Silence is also present in an act of intense looking. An artist's self-portrait is a record of this creative watchfulness. Each of the self-portraits left to us by Rembrandt, Van Gogh, and Stanley Spenser is a record of someone who is looking with total honesty into his own face. It is the record of someone who was able to sustain a paradox, for he is both seeing himself as familiar and as Other, as someone seen as if for the first time. The face is both wave and particle—both known and familiar and totally new and apart. This is why there is something so compelling about a great self-portrait. In some mysterious way, we recognize that it is the record of an artist scrutinizing his own face, rather than portraying someone else. But how are we able to recognize this?

I like to draw. As an amateur, I took a series of drawing and life classes at an art school. A beginner often makes the mistake of not really looking, for he is intent on drawing what he expects to see. But on one occasion, I had that marvelous experience of drawing with such total absorption that suddenly I did not know what was before me. A head, a hand, a foot appeared, totally alien, totally other. It was as if I was visiting Earth for the first time and had never before seen human life. I was totally absorbed by drawing a new and totally curious object, exploring it with pencil or charcoal, and interrogating what was presented to me.

Think of the total intensity of a child watching something. His eyes seem so bright. He watches in total absorption. All the time, he is internalizing and metabolizing what is going on and representing what he sees in a sort of internal code of dispositions to movement. He may twitch and squirm with the effort of watching, which can then be played out when the child performs a complex task he has learned by watching.

A further example of the power of silence comes from the theater. A good actor isn't just someone who can move and deliver lines convincingly. She can also be a strong stage presence when simply listening. Actors on stage will have heard each other's performances night after night, to the point where they becomes totally familiar. Bad actors wait, possibly even "mouthing" the lines of the actor already speaking, until they are cued to speak. A good actor, by contrast, fills the silence with something more powerful: an active listening, based on a rich inner monologue.

A famous example comes from the opening scene of Chekhov's *Three Sisters*. On Irina's saint's day, which is also the anniversary of her father's death, she and her sister Olga reminisce about the past and dream of moving from their home in the provinces back to Moscow. The third sister, Masha, absorbed in her book, says nothing. It would appear that the audience's focus should be on Olga and Irina, yet in a good production, Masha, by her refusal to join in the conversation, commands the stage. The intensity of her silence creates a tension within us. We become curious about her character; we wonder what is going on in her mind and what impact she will have when she speaks.

Chekhov knew the power of silence on the stage, as did another playwright of genius, Samuel Beckett. Much of *Krapps Last Tape* consists of the old man, Krapp, listening to tape recordings made in his youth. Even more dramatic in this respect is Beckett's television play *Eh Joe*, in which the figure of Joe is filmed in his room as he listens to a woman's voice. He makes no sounds of his own, and no one else is seen. Beckett's directions are that Joe's face should be motionless, unblinking, and impassive "except in so far as it reflects a mounting tension of *listening*" (Beckett's emphasis).

Being with the void also means holding the edge of a question and not rushing to its resolution. The physicist David Bohm expressed his misgivings at what occurred in the months following Heisenberg's discovery of quantum mechanics in 1925. Most physicists view that era as a golden age for physics. Bohm did not.

It was a period in which Schrödinger came up with his alternative wave mechanics, and he and Heisenberg discussed (along with Bohr, Born, Pauli, and others) the meaning of the new discoveries. There was a strong political motive to this, for the Copenhagen group wanted quantum theory to be accepted by the physics community in general. To this end, they pushed ahead to define the philosophy of the theory and to iron out nagging questions and difficulties.

That was where Bohm believed the fault lay. The little questions and doubts were genuine, he felt. If they had been held onto and not resolved, they would have produced an additional creative tension and a way of opening up the theory— even of seeing how it could be integrated with relativity. Instead, things were resolved too quickly, and the final Copenhagen interpretation was promulgated. The result was a form of closure that led to a dying off of creative energies. And so today, we are left with a deeply unsatisfying theory that remains unresolved, alongside relativity.

Holding onto silence and the question is characteristic of all creative artists. Federico Fellini's masterwork is, for me, *8 1/2*, a film about the making of a film. But it is also about the nature of creativity in the life of an artist. The protagonist, Guido Anselmi, is working on a new film. Months have gone by, but he is still not ready to start shooting. He has built elaborate sets, carried out many screen tests, overspent his budget, and flown in international stars who are kept waiting despite having contracts and engagements elsewhere. To make matters worse, his private life is in chaos, his marriage appears to be on the rocks, and he is taking the cure at a thermal spa.

Finally, his producer turns up and insists that shooting begin. But Guido feels that his insights have left him. His characters are no longer real or have turned into clichés. He wonders if he really has nothing left to say. In the end, he is forced to admit there is not going to be a film. The sets may as well be torn down, and the crew and actors can go home.

The film itself is a remarkable parallel to Fellini's own life. After the success of *La Dolce Vita*, some in the industry were waiting for him to fail. He had originally planned something akin to a science fiction fantasy as his next film, and, like Guido in *8 1/2*, he had constructed expensive sets. Now they were abandoned. Admittedly, he had an idea for a film, or rather, he had ideas for several scenes. He also knew there would be a protagonist played by Marcello Mastroianni. But who would he be—a theater producer, a writer, someone in a profession? Fellini had no real idea. Time was dragging on, and, as happened with Guido in the film, his own producer was pressing for action.

In the end, Fellini came close to admitting to his production director, "I don't remember the film anymore. I'm not going to make it anymore." In one interview, Fellini relates that right up to the day of filming, he had intended to back down. It was only the loyalty of his crew and his fear of letting them down that kept him going. Even on the night before the first day of shooting, he did not really know who Guido, the protagonist of his film, really was.

Of course, not all filmmakers or creative artists work this way. Suspension happened to suit Fellini's particular personality. After he had discovered Carl Jung, he remarked that although Freud forces us to think, Jung allows us to dream.

By contrast, a director such as Peter Greenaway creates a meticulous script filled with complex lists and cross-references, and even frames several of his shots with reference to compositions from paintings. His *Drowning by Numbers*, for instance, a film about male attempts to control the world though games and lists, is carefully worked out using one hundred different "numbers," which appear throughout the film in all sorts of ingenious ways (painted on cows, appearing on runners' shirts, or even spoken).

Fellini could never have worked in this way. He simply had to listen to his dreams. But it took enormous courage for him to stay with that creative uncertainty. In fact, the conclusion of his film was undecided even after the shooting ended. Fellini planned to have the last scene take place in the dining car of a train, where Guido is reconciled to the figures of his past. This sequence was actually shot. But in the end, Fellini went back to that moment when Guido sits in the car, listening to his collaborator praise him for his artistic courage in abandoning the work. While this is going on, the set is being torn down around him.

Suddenly, the figure of Maurice, the magician, appears at the window and says, "Wait, Guido! Wait! We're ready to begin." Maurice leads a procession of all the figures in the film into a circus arena. Guido then gets out of the car, picks up his megaphone, and begins to direct the action. The characters form a great dance of life as they circle the ring, with Guido shouting directions. Finally, Guido and his wife enter the circle. The film ends with the musicians being led by a small boy (Guido as a child?) who plays a fife. The boy is isolated in a single spotlight. The spotlight fades. The screen is black. The music continues. Then fades. To silence.

chapter 5

Language

In the previous chapter, we learned about the power of silence. Now we're going to see what happens when we open our mouths to speak or sing. It turns out that this is one of the most creative things we'll ever do. And if this sounds a little like hype, then I hope that you'll agree by the end of this chapter that every time we speak, we are exercising an incredibly creative power, a power that ripples through our entire society.

Speech begins early in our development. A baby's first words are generally heard about the time he is taking his first steps. From then on, the acquisition of language is exceptionally rapid, and before we know it, the young child is following us around, chattering and asking endless questions. What's more, he is inventing all sorts of fantasies and tall stories—and even making jokes. As soon as the child begins to acquire language, he is also playing with language's limits and ambiguities!

Because we can't recall our own first immersion in the world of language, we forget the way every child seems to pass effortlessly though the same process. We simply don't recognize a miracle when it's happening. Only when we go abroad and try to speak a new language do we realize just how much we must have been learning during the first years of our lives, even as we were preoccupied with walking, playing, making relationships, and finding out about the world.

Think back to the first time you learned a second language, probably at school. To begin with, you were preoccupied with memorizing verbs, figuring out the rules of grammar, and picking up vocabulary. After a few weeks, you started putting sentences together, asking for things, and replying to simple requests. Yet before speaking, you'd first have to think out what you wanted to say, forming the sentence in English and then translating it into the new language. Then one magic day, if you kept up with your studies, you suddenly found yourself speaking and thinking in the new language without having to revert to English to figure things out. You may not yet have been bilingual, but you'd have passed a major milestone on the way, and before you knew it, you had your first dream in that new language.

Previously, you had been relying on a teacher or learning stock phrases from a teach-yourself language book—securing a room at a hotel, buying a ticket at the railway station, making a purchase in a shop, or asking directions. But now you were able to say things you'd never read in a book. You discovered that you could frame thoughts, make observations, express kindness, ask about things, and argue particular points using sentences you'd never before read, spoken, or even heard. In fact, the number of possible things you could talk about—even with a relatively limited vocabulary and sticking only to the present tense—was unlimited.

And this is what we do in our native language every day of our lives. We speak and we think about things using a sort of internal language. We continue to do this for decades, but we never seem to come up against a barrier. Our ability to use language means that we have an unlimited creativity inside us.

When a young child spills her milk, she rapidly invents an elaborate story of an imaginary friend who has been naughty and tried to take the milk from her. When we adults become involved in a striking new experience—falling in love, losing a parent, having a traffic accident, visiting a new country—we still find ways of putting things into words. In other words, we are constantly discovering ways of using language to express unusual experiences, feelings, and thoughts. We sometimes say we are "stuck for words," but experience tells us that being stuck doesn't go on for long.

A good clue to the creative nature of our use of language is the way we can so easily spot when that creativity is missing. Noticing that language has been compromised should tell us something important. It's like the story of the sea captain who could sleep through the greatest of storms but woke suddenly when the ship's engines stopped. We may not always value what we have, but we certainly know when it's not there. To take another example, depression is often diagnosed by the lack of creativity in the way a person speaks. It is not only the range and choice of words a healthy person uses that communicate and convey meaning but also the tone of voice and the animation given to language. The voice of a depressed person is flat, his conversation is not full and expansive, and what he says tends to go in circles, with the same complaints being repeated over and over again. Ask him a question, and you may not get a proper answer. It is as if his creative energy is so low that he just doesn't have enough to spare to put into his speech.

By contrast, a schizophrenic may be bursting with language, to the point at which it is impossible to shut him up. But most of the time, what he says just doesn't make complete sense. Schizophrenics tend to use "word salads," stringing words together because of the associations of their sounds or shapes and the way one word triggers another. Unlike the depressed person, they can certainly talk. A hint about their inner suffering lies hidden and encoded in their bizarre utterances. Yet the overall feeling of integration in language is missing. In these individuals, a disorder of the mind has compromised the ability to convey meaning through language in ways that are clear and invite response.

Disorders of language impoverish a person's ability to communicate and convey the richness of his or her inner world. (However, it's certainly true that a depressed person is nevertheless communicating something. She is showing others that she is depressed; similarly, the language of a schizophrenic shows that his thought processes have become disordered and, to a certain degree, uncontrollable.)

Depression and schizophrenia are individual disorders that manifest themselves in an impoverishment of language use. But there are even more serious disorders that infect the whole population. They compromise full communication and even cut us off from our inner feelings. In this case, among the main culprits are politicians and military leaders, as well as social workers, some critics, and spokespeople for big businesses who resort to the use of jargon, obscure phrasing, and insincerity.

It was not simply a desire to be mealymouthed that persuaded the military to use such terms as *preemptive strike, anticipatory retaliation, extreme prejudice, terminate,* and *war games.* The use of such sanitized language is a highly effective tool for influencing and controlling others. At the height of the cold war, politicians and generals on all sides were working out scenarios in which millions of people could be wiped out. Some scientific experts even felt that the "worst-possible scenarios" of the military were cheerfully overoptimistic. Rather than whole populations being extinguished, the global cooling that would follow a nuclear war would produce a severe "winter" of several years in length, during which a great deal of plant and animal life on earth would be destroyed.

How was it possible for human beings to conceive of such horrors? Could people really bring themselves to think the unthinkable? Obviously, generals and politicians were able to spend happy weekends with their families and return to work on Monday mornings to discuss megadeaths or the strategic use of neutron bombs to wipe out all life in a city while leaving the buildings standing. The great deception was achieved through language—another example of the way in which creativity has a dark, shadow side.

How was it done? There is an interesting story of a sociology student who, wishing to understand how people could engage in such horrific acts of the imagination, took a job in the Pentagon. After several months, her supervisor noticed that her reports had become less critical of the war games mentality. When challenged, the student was shocked to realize that she had gradually begun thinking in the same way as the very people she had set out to study. Only after some hard introspection did she realize that her view of the world had slowly been changed as a result of the new language she was speaking on a daily basis.

Talking about wiping out the populations of whole cities is one thing; talking about making a "preemptive strike" is quite another. Killing human beings is something that has to be faced on the human level, but subjecting "targets" to "maximum prejudice" becomes a matter of bookkeeping.

The Nazis recognized this truth long before it was adopted into the cold war mentality. Heinrich Himmler was concerned about "the psychological problem" presented by the need for his men to remain "decent fellows" while, at the same time, shooting thousands of people. One solution was to introduce more effective technologies, such as mass gassing, for acts of "liquidation." The other was to be careful with the use of language, referring, for example, to "merchandise" or "pieces" rather than human beings and speaking of "extermination," a term generally used in connection with vermin rather than people. Another euphemism was to classify genocide as "resettlement," part of the "final solution" for various groups termed "subhuman." In this way, Himmler and his friends could be both "decent fellows" and efficient bookkeepers.

Over fifty years later, a member of a Serbian paramilitary group that had been killing civilians and burning down villages told a British reporter, "You thinking we are ruthless criminals. But we kill because of love—love of our country. . . . We are just doing ethnic cleansing, nothing else" (interview by Maeve Sheenan, *Sunday Times* [London], June 6, 1999).

Such people can maintain a genuine love of the arts, with all their aspirations for the highest values of the human spirit, while performing atrocities—even to the point of having string quartets play in concentration camps while prisoners were marched to death.

The way that the unthinkable can become the everyday is through the careful discipline of language. Specifically, this means denying certain areas of language their full flexibility of expression and using language in limited and restricted ways.

The result is that the way people feel, see, and think about things becomes similarly limited. In this fashion, men can sit together over coffee and sweet cakes to plan acts of extreme inhumanity and still remain "decent fellows."

In 1948, when George Orwell began his novel of the future, *1984*, he was writing from his experience as a writer of propaganda for the British during World War II. His novel was not so much a science fiction fantasy of a world to come as a direct projection of what he saw happening in the Britain of 1948— hence, the reversal of the date in the title. Most perceptively, he realized that to control a population, you first have to control the language. In *1984*, a government ministry is given the responsibility of creating and monitoring "Newspeak," a way of controlling thought processes by restricting the range of things people can say.

One of the characters, Syme, is working on the eleventh edition of the Newspeak Directory. He extols his work ("It's a beautiful thing, the destruction of words") and boasts that he is eliminating hundreds of words a day, "cutting language down to the bone."

Traditional English allows for nuance and shades of meaning, but Newspeak will make "thoughtcrime" an impossibility, for there will be no words in which to express a subversive thought. And when each word is rigorously defined and shorn of its shades of meaning, the range of human consciousness will become smaller and smaller. Thus, for example, the word *free* is restricted to such meanings as "a field is *free* from weeds" or "a dog is *free* from fleas." There are no longer words in which to express such concepts as political or academic freedom or the ability to think in a free way. As Syme puts it, "The Revolution will be complete when the language is perfect."[1]

---

[1] During World War II, Orwell (Eric Blair) became involved in British propaganda, broadcast to India from the BBC. In *1984*, the ultimate threat is to be taken to Room 101, which contains "the worst thing in the world." It is no coincidence that Orwell's broadcasts were made from Room 101 in the basement of BBC's broadcasting house!

Compare the paring down of language so as to reduce the ability for subtle or subversive thought with the vastly extended vocabulary of Siberian shamans. The Yakult of Siberia, for example, use over twice as many words as the average European; medicine people from Native American groups also have extended vocabularies. An expanded vocabulary is evidence of access to an expanded reality and the need to discriminate subtly different states of consciousness and reflect on encounters with energies and powers of other worlds.

We should not be too proud of ourselves if we feel we escaped Orwell's *1984* unscathed. When the book was published, social commentators were preoccupied with the future of technology and a world in which an entire population could be monitored by Big Brother. What they had forgotten was the more significant message of the book: the way the thought and behavior of entire populations can be controlled through the restriction of language. This very form of control has been going on in much of the world since *1984* was written.

Fortunately for the rest of us, the language of the military never really caught on. Most of us were too horrified with the prospect of a nuclear war to fall into the trap of this particular form of doublespeak. However, politicians have long been at work within the language, changing the way certain words and phrases function in order to misdirect without being accused of downright lying. (One of the most staggering examples of this occurred while this book was being written when an American president sought to bend language in such a way that he could openly deny his sexual relationship with a young woman! At one point, when questioned by a grand jury on this matter, he replied, "It depends on the meaning of 'is.'")

In the past few decades, we have been made aware of the subtle ways language positions women. Thanks to the writings of feminists, we have recognized the ways we had been using words to control, to foster prejudice, and to limit people's basic freedoms. Our increased sensitivity to language would change our perceptions of the world in subtle ways.

The approach of feminists was extended to increase our sensitivity about language in the context of a much wider variety of groups. But within this attempt to purify language was a certain element that chose to do things in argumentative and proscriptive ways. Inevitably, the whole debate became polarized. Those who supported a new terminology were portrayed in emotive terms with words such as *strident* and *PC*. Those who opposed were considered to be conservative, right-wing, fascist, homophobic, or racist. Inevitably, ordinary people trying to go about their lives as best they could found themselves caught up in the whole debate. For fear of offending, they were not even sure what sorts of words to use anymore when talking to others. They felt constrained and embarrassed, and many offices circulated a list (frequently revised) of acceptable forms of address and ways of referring to minorities. (Incidentally, a number of people have told me they feel irritated at being lumped together within the one melting pot as "persons of color," a term, they point out, that includes everyone who is not colored pink.)

The problem with such terms is that they are overly general, and they often bend language out of shape in an effort to be tactful and avoid offense. The writer Andre Dubus, in his essay "Light of the Long Night" about the car accident that broke his legs and right hip, ends by saying, "I am forever a cripple, but I am alive, and I am a father and a husband, and in 1987 I am sitting in the sunlight of June and writing this" (*Broken Vessels*, 1991). When he was interviewed for the CBC program *Writers and Company*, Eleanor Wachtel asked him why he preferred the word *crippled* to *differently abled*. Dubus replied, "I think it's because I'm a writer; I don't like euphemisms; I don't like words designed to cheer me up." He referred to his eldest son telling him, "You're not handicapped, you're physically challenged. It's a new phrase," to which Dubus shot back, "*You're* physically challenged; you're breathing hard from walking the stairs. I'm crippled."

In relating this story, my point is that overgeneralizations can be taken for euphemisms. Some people would find it offensive to be called "crippled," "blind," or "deaf"; others would be equally offended by being referred to as "differently abled" or "physically challenged." Language asks us to be ever alert, ever specific, ever perceptive. But having started from the very best of motives, we find that we are increasingly restricted in the free use of language. As a result, our worldview has become somewhat fragmented and uncomfortable. Parents who find their children a pain in the neck employ a nanny so they only need to spend "quality time" with their offspring. For my part, I'm no longer asked to explain things or tell a story; instead, I have to "share" with others. And woe betide me if I ever have a memory lapse because pretty soon people will take a second look at me and refer to my lapse as a "senior moment." Probably the only people who can save us at this juncture are the comedians and satirists whose daily traffic is language.

At the height of the civil rights movement in the United States, when busing had become highly politicized, the stand-up comedian Lenny Bruce noticed black comedian and activist Dick Gregory walking into the club where he was performing. Bruce at once drew attention to the "nigger" who had just walked in. The audience responded with a gasp of shock and then laughed in an embarrassed way. But Bruce had only just begun. He also referred to the number of "kikes," "micks," "spics," "guineas," "greaseballs," and "Polacks" in the audience. By openly using such terms, he forced members of the audience to confront some of their own prejudices.

The laughter that night was far from good-natured. Gregory himself was angry until he realized the way Bruce was working his audience. People were alternatively embarrassed, shocked, ingratiating, and hostile. Their laughter betrayed a whole gamut of painful and generally unacknowledged feelings. Bruce was working at the level of language itself by openly using words that most liberal people would not admit to thinking, let alone using. What he achieved was far more powerful than any social or political essay. The very currency of the words of prejudice had been changed, and no one present or anyone who has since heard the recording of his performance would ever again be able to use those words in an unthinking way.

Bruce was a clown in the deepest sense of the word. In many societies, the clown is a far different character than the funny personality who performs slapstick in a wig and red nose. Although the clown may, indeed, make people laugh, he or she is also questioning and transcending the boundaries that society assumes. Clowns terrify and disturb. They have an important role in some Native American religious ceremonies, where their actions, to an outsider, can appear blasphemous and obscene. Lenny Bruce was such a clown for America in the early 1960s.[2]

It would have been easier to find a softer, warmer example for this chapter. But being warm and appealing was not Bruce's role in life. He was arrested in San Francisco under Section 311.6 of the California Penal Code for issuing an obscene word. Many people appeared as witnesses for his defense. One of these, the columnist and critic Ralph J. Gleason, argued that Bruce, as a performer, was exploring "a semantic proposition" that "words have been given in our society, almost a magical meaning that has no relation to the facts" (Gleason speaking at Bruce's trial, as quoted in Lenny Bruce, *How to Talk Dirty and Influence People*, 1965). Bruce's act would demonstrate that there is no harm inherent in the words themselves. For him, the obscenity lay not in the words but in society's hypocrisy, and his specific targets were those who held and abused power.

------

[2]Recently, I saw a contemporary version of Bruce's attack on language and thought offered by the actor Jack Klaff. During a science conference, he gave a "lecture" in the role of a certain Professor Singleton-Guinness, who had been awarded the Nobel Prize for reductionism in science. Klaff performed his lecture in a straight-faced, academic way, making reference to slides and so on. Then, in the midst of the audience's laughter, he suddenly asked them to consider whether results of the medical experiments performed in the Nazi concentration camps should be used by the scientific community. "Was it good science? Were the data useful to medical researchers?" he asked. The audience became hushed, and several people were so offended that they walked out. As we later learned, his most outrageous statements were not part of a fictional script but direct quotes from living scientists. Klaff felt he had to go over the top to make the audience realize the dangers inherent in a certain scientific approach.

Bruce was a comedian willing to defend language to his last breath. He was fighting at the barricades alongside poets and writers. In this, he exemplified a story from ancient China. During a troubled period, the emperor summoned Confucius and asked him what action he could take. "First purify the language," Confucius said.

This is it in a nutshell. Language is the repository of meaning for the whole of a society. Language enfolds the way we see the world. It is present when we encounter other people, and it influences what we think about ourselves. When language becomes devalued, restricted, and debased, then so do our thinking and our sensibilities. Stanley Kubrick's film *Full Metal Jacket* showed the way a boot camp sergeant verbally abused his men to harden their sensibilities in preparation for a tour in Vietnam.

Enter any profession—law, medicine, science—and you'll find that, though you won't be verbally abused, you will nonetheless go though a form of conditioning via language. In these cases, the conditioning is not designed to restrict how we think but to open our sensitivities to nuances most people never see and to learn ways of communicating these subtleties to other members of the profession.

The way a society's worldview is related to the language its people speak was studied by two linguists, Edward Sapir and Benjamin Lee Whorf. Whorf had spent time with the Hopi Indians of North America. He noticed that, for them, space and time are not separate concepts but are unified into a space-time notion, not unlike that of Einstein's relativity. This same space-time concept was also present within their language. What is more, their language differentiated between things that exist and things that are on the horizon of coming into existence. We can, of course, make such distinctions in English, but it takes a bit of effort, and we have to talk around the point and use metaphors and analogies. For the Hopi, however, the perception and the language fit perfectly into their overall worldview.

We do not need an academic linguist to tell us that language and worldview are deeply linked. For most of us, this observation is a matter of common sense. Many Native American elders explain that the only way to understand their philosophy at a deep level is through a knowledge of their language. And when they speak of an indigenous science, they point out that their own particular scientific tool for dealing with the world is language itself.

The Blackfoot, for example, speak one of the Algonquian languages, which are all strongly verb based. In fact, verbs are the most important things in their language, with many nouns being derived from verbs. Verbs represent processes and movements, and not surprisingly, the Blackfoot see the world as flux and process. In the Blackfoot universe, nothing is static. Nothing lasts forever. All is in movement and transformation. Contrast their language with European languages, such as English, in which nouns plays a stronger role. "The cat chases the mouse" suggests two distinct entities, a cat and a mouse, linked together though a verb, the action of chasing. In Blackfoot language, what would be primary would be the movement, the chase itself, with the cat and mouse being secondary to this immediate experience of the flux and the process of emerging out of it.

Curiously enough, a similar point was made by the physicist David Bohm, who, at the time, knew nothing about the Algonquian languages. Bohm and the physicist Neils Bohr before him had argued that although scientific equations are written down in formal mathematics, their meanings and applications have to be discussed in ordinary language (English or German or whatever) whenever scientists meet. But these languages have developed though our interaction with the large-scale world. They contain our experiences and assumptions about space, time, causality, and so on. And when physicists speak about their subject, they continue to embrace these assumptions in largely unconscious ways.

Such notions do not apply at the quantum level—a world of flux and transition. In such a world, there are no true objects possessing intrinsic properties. Rather, there are processes, transformations, and symmetries. Bohr argued that the hidden "classical" assumptions within language place a strict limit on what we can say about the quantum world: "We are suspended in language so that we don't know which way is up and which is down" (Bohr, quoted in Aage Peterson, *Quantum Physics and the Philosophical Tradition*, 1968). As soon as we begin to ask questions about the quantum world or construct models of the subatomic process, we import classical (large-scale) concepts into the subatomic domain—concepts that have no business being there. The result is paradox and confusion and, in Bohr's opinion, a limit on how far we can go in understanding the quantum domain in terms of ordinary thought and language.

Bohm was less pessimistic on this point. He agreed with Bohr that our everyday languages are replete with classical concepts. Moreover, their subject-object division and extensive use of nouns cause us to divide the world into objects in interaction. This totally goes against the whole approach of quantum theory, which deals in quantum wholeness. Nevertheless, Bohm believed that human beings could respond creatively to this situation by developing a new language, which he called "the rheomode." This language would be largely verb based and deal directly in process, transformation, and flow. It would be ideally adapted to discussions of quantum theory.

As a practical tool, Bohm's rheomode was never really tried out. Most linguists feel it is not possible to create a working artificial language in that way. However, shortly before his death, Bohm met with Algonquian-speaking elders and discovered that their language had much in common with his dream for the rheomode. Moreover, their worldview and philosophy were very much in sympathy with those demanded by quantum theory. It will be interesting to see what happens when native Algonquian speakers also become trained as theoretical physicists!

As we have seen, language is deeply tied to the way we see and act toward the world. When language changes, when it becomes rigid and restricted, or when it suddenly expands with new energy, our vision of the world also changes. Language is a repository of meaning for the whole of society. It is the vast pool at which we all drink. It is a pool of meaning to which we all contribute.

Each one of us has an obligation to care for this shared pool of meaning, to keep it pure, maintain its flexibility of form, and guard against its pollution. I don't mean we should avoid making grammatical mistakes when we speak or set up a committee for the purification of language. Such approaches would be both impractical and dangerous, for language is a living thing, and its structure is in constant transformation. No, I mean that we should be giving attention to the whole context in which language is used, to what is being said, and to the way the form of what is said matches its context. In particular, we should be alert to those occasions on which language is used in dishonest and misleading ways.

We have a duty to individual words, which, from time to time, require creative renewal and revival. Poets work with language in such a way: They help to polish words, just as jewelers polish rare stones. They place words in their proper settings and thereby enrich them. Comedians also work with words, contrasting them and emphasizing paradox and surprise. Orators use words to seduce and inspire. Therapists watch the way words are used as clues to the way people are blocked and traumatized in their feelings and thought. Above all, when it comes to the vitality of language, the ultimate responsibility for creative renewal comes down to each one of us.

Language and Perception

Linguists do not all agree as to the precise nature of the proposals being made by Whorf and Sapir. Were they suggesting that the language we speak determines the world we see? Did they contend that the social and natural environment in which we live determines language? Or did they really mean something in-between?

If, like the Inuit, we grow up using many different words to describe snow, will we have a far more sensitive perception of snow than non-Inuit speakers? Or, by contrast, if we grow up in an environment in which our very survival depends upon understanding the properties of snow at different temperatures and degrees of compaction, will we inevitably develop the appropriate facility within language for such important distinctions? Surely, the answers to these two questions are inseparable. When biologists, physicists, composers, doctors, or lawyers want to make subtle distinctions of classification, they invent or adopt words accordingly. Part of a person's apprenticeship in one of these fields involves learning how to use a new terminology in the correct way. In turn, by exploiting this extended system of classification, the expert is able to make distinctions that lie beyond the layperson. But of course, Whorf and Sapir were talking about far more than the adoption of particular nouns. Their investigations were about the whole structure and use of a language and the way it is used in different ways by different societies.

Why do a hunting people like the Blackfoot have a complex verbal structure to their language but a settled people like the Iroquois have a language so rich in kinship terms? In both cases, the language reflects the group's traditional lifestyle. The Blackfoot's language is the product of the flux in which they once lived, just as the language of the Iroquois is the product of a lifestyle centered on village life and traditional ways that made the people sensitive to the nuances of complex family relationships. Indeed, native Iroquois speakers point out that it would be impossible to fully understand the richness of their language unless you had grown up learning about life in the community.

The small Italian village in which I now live has a similarly complex web of relationships. Families have lived there for hundreds of years, and it takes quite a time to discover just who is related to whom. One person who married into the village told me that for the first few years, she had to be very careful in speaking of others because the person she was talking about may well have been the second or third cousin of her neighbor! Obviously, knowing the various ways in which people are interrelated is enormously important, yet the language never developed the nuances of the Iroquois. Indeed, the word *nipote* includes both nephew, niece, grandson, and granddaughter; the ambiguity can only be removed by understanding the context in which it is used. Likewise, people may refer to a *cousin,* using this word to include not only an actual cousin but also a nephew or brother-in-law. It seems that the precise relationship is less important than the fact that the person is gathered in as a family member.

The relationship between language, society, and perception is far from clear-cut. But evidence does suggest that the language we speak and the way we look at the world are related. In one series of experiments, for example, rapid eye movements were monitored while subjects looked at a series of photographs of groups of people. By observing the areas the eye had interrogated, it was possible to get a general idea of what was being "seen." Sure enough, what people "saw" in the photographs was determined by what had previously been said to them. In other words, the verbal context in which a scene is presented to someone determines the particular intentionality of perception.

These and related experiments show the complex ways in which language and vision affect each other. They underline the fact that seeing is a creative act. Language and perception are inseparable, and although this chapter is primarily about language, it will be useful to devote a little time to the creative nature of seeing. For seeing the world around us and reading a poem are, in many ways, related activities. They are both creative and interpretative processes.

A naive theory of language speaks in terms of packages of information being transferred from speaker to listener. A more satisfying theory holds that a conversation is a mutually creative act in which both speaker and listener are building up and furnishing their own mental spaces until they finally take on a communal reality. Something very similar happens with perception. A naive theory of perception would suggest that, as with a camera, data are gathered by the eye and transferred via the optic nerve to the brain, where some sort of electrochemical "photograph" of the scene is produced. (Of course, this begs the question, for if the brain contains an "image" of the outside world, then who or what is, in turn, viewing this image?)

The actual state of affairs is far more complex and creative. It is true that the optical system of the eye produces an image on the retina, which, though a series of electrochemical processes, is then converted into signals along the optic nerve. But what happens when these signals travel into the brain toward the visual cortex is more interesting. To begin with, the traffic is not one-way. As incoming data move toward the brain, they are met by a stream of data traveling down from the visual cortex. All along the optic nerve, incoming data are being analyzed, compared, and processed, using signals from the higher centers of the brain. Then, when the data reach the cortex, they do not end up in one localized region but are routed to several quite different areas, each of which is responsible for different forms of visual processing. Finally, this complex series of cerebral events becomes the subjective and totally effortless act of seeing.

Just as we create a mental space during a conversation, the brain constantly makes hypotheses about the visual world and seeks to verify them by gathering even more data. Suppose you are out for a walk in the country with someone who is speaking about a mutual friend, another outdoor lover. Your brain may form the hypothesis that such a person is present. Signals are then sent to the muscles around the eyes, which direct your vision to hunt around the scene and watch for something that looks like a face or body. These motions of the eyes are extremely rapid, with the result that there is a constant mass of data flowing toward the brain, including information about parts of the undergrowth that could conceal a face. And if something that could well be a face is seen in the shadow under a tree, then the eyes will at once scan around that general area looking for more information.

Within the visual cortex, these data are being analyzed in a variety of ways. Parts of the cortex are looking for contours, that is, for the outlines of objects and shadows. Others are seeking out movement, fields of color, and so on. As fast as the information is coming into the brain to confirm or deny a visual hypothesis, the brain is making yet more hypotheses, redirecting the eyes to check out various regions of the scene, and actively analyzing incoming data.

Finally, a face is recognized as coming out of a shady part of the woods. Immediately, the eyes are directed toward that region, so that the image of the face falls on the retina's yellow spot, that region containing the densest collection of receptors. In turn, data about the face reach parts of the visual cortex that are particularly rich in the ability to discriminate and process data. Much like the zoom lens on a camera, the brain is able to magnify and gather a much higher density of information in a particular region of the visual field.

Finally, the face is recognized as that of one's wife, not the friend who was being discussed. Now a new set of hypotheses and visual questions are set into motion, and the eye and brain move to scan the scene afresh and seek out new data.

Estimates are difficult to make in this regard, but it may well be that around 50 percent of what we see comes from the scene outside and 50 percent comes from what is already in the brain—that is, from what is known about the world, the nature of movement and contours, the colors of objects, the way light and shadows behave, and so forth. In other words, we are not so much seeing a raw, unmediated world but something closer to a scientific theory of the world—the end product of a long series of hypotheses that have finally been verified though visual experiments. And amazingly, all this happens in the blink of an eye.

Vision is intentional. What we see is not so much given as built out of what we know, what has been programmed innately into our brains, and what we are actively gathering from the world outside. If we change a conversation or a context, we look at the world around us in a new way, noticing visual nuances that had previously been ignored.

Every action of opening the eyes is creative. The process involves the same creativity used by generations of painters who understand the way human vision works and employ this knowledge to make coded marks, gestures, fields of color, and contours that evoke in the viewer a rich visual experience.

Some painters, such as the Pre-Raphaelites and the Magic Realists, used this encoding to give us a dramatic or heightened illusion of reality in their paintings. Others, such as Velásquez and Rembrandt, understood the way that thought and feeling are related to perception, and their portraits give us deep insights into the inner lives of their subjects. Yet others have used the visual codes to deconstruct our acts of seeing, as, for example, with cubism.

Mental Spaces

We can use the common pool of language in many different ways. One way is called "phatic communication" and is the sort of thing we do when we meet a neighbor and exchange greetings or a few remarks about the weather. The purpose of phatic communication is not so much to convey information—everyone knows what the weather's like at the moment—but to cement friendship and keep the contact alive.[3]

At the other end of the spectrum, language can be used to convey a series of facts. A television weather reporter gives the predicted temperatures across the country for the next twenty-four hours. Someone else reads the stock market reports. Here, language is being used in a very straightforward way to transmit packets of information from one person to another, much like two computers transferring files over the Internet. However, unlike phatic communication, merely exchanging a string of facts involves little human warmth or engagement.

Most of the time, we occupy a rich middle terrain between these two extremes. This is where our essential creativity lies. If all we ever did when we met other people was "transfer

---

[3]Phatic communication is a vital element in the village in which I now live, just as it is in many Native American communities. Each day, I go on the *giro* (a walk around the village), stopping in at the shops and at the bar for a coffee. To walk around the village actually takes five minutes, but the giro can take well over an hour. It is a time to greet people, talk about food and the weather, exchange a little gossip, and renew connections. It is a flow of warm conversation, confirmation of one's position, and an interchange of casual information that winds its way around the village each day. Everyone taps into this pool of shared companionship.

files," we'd end up lecturing them in a neutral, impersonal way and never form friendships. But if we only used phatic communication, we'd start off feeling warm and close but never go any further in getting to know others; we also wouldn't be able to organize any collaborative activity.

What we do instead is remain friendly, give the other person lots of space in which to join in, reply, maneuver, and, at the same time, exchange some information—opinions, needs, wishes, queries, and so on. To do this, we have to tap the vast resources of language.

As we speak, we try to keep our own options slightly open because we also want to know what the other person thinks and see how he or she reacts to what we're saying. After all, nothing is worse than a bore with fixed opinions who drones on and on and causes all the others in the room to keep looking at their watches.

In a creative conversation, we find ourselves dealing in shades of meaning, ambiguity, and even some paradox. We're no longer employing packets of cut-and-dried facts to be exchanged back and forth; now we're working with opinions, nuances, and possibilities whose full meaning and significance depend upon the overall context. This context is not fixed once and for all but changes even as we speak. The whole field of meaning is itself changing and fluid. When we talk in a casual way, we don't begin by defining every concept and trying to pin down every idea. Rather, we allow things to float a little and find their own levels.

The French linguist Fauconnier compared the way conversations work to building what he calls a "mental space." Suppose someone expresses an opinion or discusses an idea with you. As you listen, you begin to form an overall notion in your head about what that person means. In turn, the replies you give help the first speaker to realize how well the particular point is getting across. Rather than transferring files of neutral, unambiguous data, both parties to the conversation are building and furnishing their respective mental spaces with ideas, concepts, feelings, and attitudes.

In such a conversation, there will be times when these spaces are not congruent because meanings are misinterpreted or contexts misunderstood. Normally, we are able to handle these discontinuities easily and on the fly. But in extreme cases, they can lead to a serious breakdown in communication between two people. This sort of situation is encountered by a marriage counselor whose clients have been bickering for years over a series of apparently small yet irritating points. Therapists would say that is because the husband and wife are not "hearing" each other; they long ago built up a series of mental spaces and invested them with considerable emotional baggage. The problem now is that these spaces are no longer congruent, so when one person speaks, the other forces the remarks into his or her own mental space, badly distorting them in the process. But with the help of a counselor, their mental spaces will shift and begin to free up if they really make an effort to see their spouse's point of view. In this way, creativity returns to a dialogue and to a relationship; a new mental space is built in which meanings begin to move in subtle ways according to the new context.

A conversation becomes a dance of meaning. It is a symphony of language in which each person is constantly creating, like an artist or a composer, harmonies of meaning and context. This symphony takes place between people, inside their heads, and within language itself, which is the common property of the whole of society.

The German artist Joseph Beuys claimed that everyone is an artist. He also introduced the concept of "living sculpture"— the notion that our thoughtful and creative social actions form a sculpture or a creative movement within society. As we have just seen, having a conversation in which both parties are open, respectful, warm, and yet able to explore ranges of meaning is, in itself, a piece of living sculpture. Such an exchange resonates throughout the whole of society, for creative conversations are constantly renewing the pool of language, restoring its vigor, and counteracting the rigid tendencies of less creative conversations.

The significance of a conversation can be taken even further. Decades ago, neuroscientists believed that although the brain develops in the womb and during early infancy, its structure is more or less fixed after the first years of life. They felt that it is always possible to learn new facts and deposit them in one's memory but that the brain itself is hardwired like a computer early on; thereafter, nothing will modify its internal structure. But then, neuroscientists such as Eric Kandel, who was working on the nerve structures of sea slugs, noticed that the fine structure of connections can be modified even in adult animals. Though it is still true that the gross structure of the brain is more or less fixed, all its fine wiring—the delicate connections between neurons—is constantly changing. So when we learn something new, it is not just a matter of some new file of information being stored in our memory but of the very connectivity of the brain; the way it processes information, experiences the world, and has feelings or desires also changes.

This means that when we talk, our symphony of meaning is operating on several levels at once. At one level, our abstract mental spaces are being visited by interior decorators and modified by builders. At another, language itself is being renewed and extended in such a way that the effects will permeate society. And at yet another level, the physical structure of our brains is changing. We are literally becoming different people each time we speak.

A chance meeting or a casual conversation thus becomes enormously important. It suggests that every moment of the day, we are on the threshold of creative change. We may not realize it at the time, but when we talk, we are constantly remaking ourselves and, as living sculptors, helping to transform the society in which we live. The key and the most creative thing we can possibly do is to stay alert when we speak and remember that, in each moment, we stand on the brink of a new life. What we say to another person could have creative repercussions that ripple out in space and time. Conversely, if our words are harsh and cruel, they could cause the creativity around us and within us to wither and die.

The writer Dostoyevsky had a horrific personal experience that he worked into one of his novels, *The Idiot*. Arrested on political grounds, he was told that he was going to be executed. The writer was taken from his cell and marched in front of a firing squad. He was then blindfolded, and the order to fire was given. But nothing happened. The whole procedure had been a mental torture designed to break the prisoner.

In *The Idiot*, Dostoyevsky relates a similar occurrence from the perspective of one of his characters. On the way to the execution, the condemned man notices with exceptional vividness every nuance of the day. How wonderful life appears! He would give anything just to have an extra minute of life. How dull and stupid he has been never to have noticed the world this way before. If only his execution could be stayed, he would live every moment of the rest of his life with intensity. But in truth, Dostoyevsky points out, the reprieved prisoner would return to his old, stupid, dull, and habitual existence within a matter of hours. The immediacy of the world would fade, the sense of newness would be gone, and he would soon find himself complaining about his lot, just as before.

We are always before the firing squad. Every conversation holds within itself the promise of a new beginning. But no matter how vivid the potential for change, it is all too easy to slip back into old habits. The theme of this book is that the habitual way of looking at things is, in fact, an illusion. It is not a true reflection of the world but a fallacy. All of the cosmos, from stars to elementary particles, is essentially creative, and each time we speak or see or listen, we are being creative. The trick is to stay awake, to remember who we are, and to avoid the dulling drug of habit.

Toward the end of his professional career, Sigmund Freud introduced the idea of the death wish. For much of his life, he had explored the power of the sexual instinct, but now he suggested that its direct opposite was also present in human beings. Just as there is a force toward life, there is also the seduction of death and a return to the inanimate—to a state without choice, potentiality, and the constant possibility of transformation. Freud saw this as the essential condition of inanimate matter. But in this book, I have proposed that all matter is alive and carries within itself the potential for transformation. So the death wish is really the desire for some illusion of stasis. It is that deadening cloud that envelops us and causes us to forget, to sleep, and to deny our essential creativity and the possibility of change. If we do stay awake and alert to the possibilities of the new within each greeting and each conversation, our brains will constantly change, adjust, and transform.

This alert transformation can continue into old age. A few years ago, I spent an exciting afternoon with the composer Michael Tippet, who was then ninety-one. He was totally alert, perceptive, and creative in the way he parried many of my questions. Thirty years earlier, I'd visited the philosopher Bertrand Russell, who, at ninety, was an active worker for world peace, writing to the heads of state in many countries and pushing ahead his war crimes commission.

Today, it has become something of a fashion to describe any elderly person who appears confused and forgetful as suffering from Alzheimer's disease. That blanket description covers a multitude of conditions. Some old people do experience physical degeneration of the brain. Others may be suffering the effects of poor diet, restricted circulation to the brain, or an overuse of sleeping pills and tranquilizers. Many more will be victims of restricted stimuli. No longer able to get out of the house, extremely deaf, deprived of human contact, or unable to see well enough to read, their world slowly collapses in on them. They are no longer capable of engaging in life in open and creative ways. Starved for excitement and interest, their faculties begin to degenerate. And the more this happens, the less likely other people are to spend time with them. But something as simple as a pet, a weekly drive in a car, or a home visitor can quickly turn the tide and restore such people as valuable members of society, with a repository of memory and wisdom to offer.

Apart from disease and the side effects of poverty, there is no reason we should fear old age. Michelangelo was carving marble and designing architecture through his eighties. Picasso remained vigorous into extreme old age. When Matisse could no longer hold a paintbrush, he developed a radically new approach, designing remarkable and colorful abstracts by cutting out shapes in colored paper. The vigorous third period of Verdi's composing life, which included operas such as *Falstaff* and *Othello*, began when he was seventy-three. In old age, with his sight almost destroyed by cataracts, Monet began to paint in a totally new way, producing vast canvases so big that the wall of his studio had to be opened in order to get them out. These enormous works show the surfaces of lily ponds in which sky, water, willows, lilies, and reflections float across the canvas. In fact, they are so big that one does not so much stand and look at them as walk back and forth to explore them.

If you enter a Native American community, you will find that the greatest respect is given to the elders. They are the repositories of history, traditional stories, and deep wisdom. It is a great gift for a young child to grow up close to a grandparent or great-grandparent and begin to learn something of the past and of the experiences of decades. (It is also a gift to an elder to have the privilege of passing on the experience of a life.) Rather than being relegated to retirement centers or nursing homes, such people have entered a phase of existence that should not be lost to society at large. They are the great oaks of our modern world. They are there to remind us to listen, to become aware of our roots, and to ensure that we too pass on something to future generations.

Virtual Reality

Language allows us to live in a wider reality, a reality that extends beyond our immediate body senses into the world of the imagination. (In Chapter 7, we shall see that words and language are indeed embodied, so this virtual reality of language extends from the abstract right into the matter of the body.) Those who have a pet will probably believe that animals do have a sort of imagination and can entertain fleeting images of things that are not there. But language gives us the ability to build up within the mind whole worlds and to furnish and transform them. What's more, these worlds can be communicated and shared. And so, we can live in books to the point at which the characters become more real than our old friends.

There is an amusing anecdote about past-lives regression. Some people believe that they once lived as famous characters of history or witnessed key moments, such as the storming of the Bastille or the landing of the Pilgrims. It also turns out that some people sincerely believe that, in their past lives, they were Jane Eyre or David Copperfield—characters out of fiction! Such characters have become so real in the public imagination that they have their own history of existence in our world.

A similar phenomenon occurs for the aficionados of TV soap operas. In Britain, the most popular soap is *Coronation Street*. When the plot line involved the wrongful jailing of one of the characters, public outrage reached such a level that one MP actually posed a question in the House of Parliament about a possible miscarriage of justice—all for a fictional court case involving a fictional character![4]

We may laugh at the sort of people who get caught up in fiction to this extent. Yet the fact that they do tells us about the creative power of human beings to build entire universes of the imagination, to share them across vast sections of society, and to inhabit them for a part of each day. Much is boasted today about computer-generated "virtual reality." But is it really that new and dramatic? Millennia ago, the stories told around the fire at night (the ancestors of virtual reality) had even more powerful repercussions for society.

---

[4]In the British parliamentary system, a period known as "question time" is set aside each week. During this period, any member can ask a formal question of a government minister. (Sometimes, detailed notice of the question is given beforehand.) Some questions are designed to embarrass the government and raise doubts about official actions and current policy. Others, asked by members friendly to those in power, allow ministers to expound on recent successes or define new interpretations of policy. Probably no one imagined the period would ever be used to plea the fate of a character in a soap opera!

Language has yet another powerful potential in its further abstraction into algebra and other branches of mathematics. In part, the underlying structure of language involves sets of logical transformations and connections. By codifying the rules and substituting abstract symbols for them, we can begin to move from everyday spoken language into formal logic and eventually into abstract mathematics. In the early Middle Ages, mathematics involved a combination of numbers, symbols, and ordinary words. Only gradually did it become totally abstract, as it is today.

Numbers and mathematics allow us to perform further acts of abstraction. The mind of a scientist can create a world of atoms, elementary particles, galaxies, and stars or even an exploding universe. Or it can move into a world that is entirely abstract and with no apparent connection to anything material. The world of pure mathematics is a universe of pure thought. It is something created for joy and for its aesthetic appeal.

The ability to abstract and make real within the imagination has produced all our technology. New technology requires a leap of the imagination. It is born of the ability to posit something within the mind, consider it, manipulate and transform it like a real object, and finally realize it within the material world in steel, concrete, plastic, electronics, chemicals, and so forth. In turn, new objects in the material world can be further abstracted and played with in the mind to suggest new technologies.

Let us take a particular example of the power of abstraction. Before the Industrial Revolution, people created intricate patterns for carpets and other fabrics on looms in their homes. Then industrialists realized that the various operations of weaving could be mechanized, with instructions expressed as a series of holes punched into a card fed into a Jacquard loom. The principle was something akin to the pattern of holes used in a player piano.

These cards gave Charles Babbage the idea of building a calculating engine that would weave numbers instead of cloth. Blaise Pascal had once invented a machine to perform the repeated calculations needed for his income tax account, and the philosopher and mathematician Wilhelm Leibnitz had developed his stepped reckoner. But until Babbage's engine, all such devices required a human operator to reset the machine at each stage of the calculation. Babbage's flash of insight, derived from his observation of complex weaving, was that such a machine could be "programmed" using punched cards so that a whole sequence of calculations could be automatically set into operation.

Like a modern computer, Babbage's engine was to have a "store" for data and a "mill" for number crunching. Working with Ada Lovelace, the daughter of Lord Byron, he wrote the world's first computer programs, using the results from one calculation as the data for the next. Since each specific area of a card either contained a hole or did not, the use of punched cards by Babbage and Lovelace was exactly equivalent to a binary code.

Babbage attempted to build a prototype of his mechanical computing engine but was constantly frustrated by the lack of high-quality craftsmanship in those working for him. In fact, the full potential of Babbage's engine had to wait until the age of electronics and the coming of another thinker of genius, Alan Turing. In an effort to break the German Enigma code during World War II, Turing designed COLOSSUS, an early computer. Although Turing's computer used electronic tubes and electrical currents instead of cogs and wheels, it operated according to exactly the principles Baggage had proposed. Instead of punched cards, it employed continuous rolls of paper tape punched with a binary code. Around the same time, the University of Pennsylvania built ENIAC, with the help of John von Newmann, to produce accurate bombing tables.

Soon, electronic valves gave way to semiconductors, semiconductors to printed circuits, and so on to ever more miniaturized circuitry. As computers became faster and more powerful, it became feasible to create imaginary worlds—virtual realities. Moreover, computers began to communicate with each other at high speeds over the Internet, to the point where a cyberspace has become delocalized across the globe.

Today, our senses can be plugged directly into these spaces, and we can see and "feel" what it is like to negotiate in other worlds, some real and some the abstract products of pure imagination. The first chips have already been inserted into volunteers, enabling new sorts of connections to be made directly to computers. Moreover, it is now technically possible for us to receive data from remote sensors on the Moon or Mars and have the direct experience of driving and operating machinery on another planet. And all this derived from the abstraction of weaving!

But why stop at real planets? Why not create worlds of our own? Why not populate and live in them? Such worlds could be shared by large numbers of people. Soon, it may even be possible to have a shared consciousness—all made possible by the power potential in language.

Maybe even language itself will change. Earlier, we saw that Algonquian speakers of North America use a language perfectly adapted for their world of process and flow. By contrast, our own "noun-based" language, a language that has served us well for thousands of years, is not well adapted to the quantum world, for this is a world less about objects in interaction than about transformation, superposition, flux, and the inseparability of observer and observed.

But it is wishful thinking to suppose that physicists or philosophers can change the language we all speak. Leibnitz once tried it, proposing a philosopher's language that would be free from ambiguity and in which every term would be precisely defined. But language cannot be pinned down in that manner. It has an inherent richness and flexibility. Much of its poetic power lies within ambiguity and the ability to exploit metaphor.

However, what will happen as more and more of us interact via computers? What will happen when we spend increasing amounts of time working with a combination of words, icons, muscular movement of the mouse, virtual presence, and the tactile feedback qualities of a computer glove? What evolves could well be our new language, a language that has a very different quality than that of a conversation or a printed book. It may be that, in response to all this change, the way we speak will begin to be transformed in very subtle ways.

Equally well, all this could be a fantasy. The social impact of the computer, the Internet, and virtual reality may have been vastly overestimated. But at least we have been able to indulge our fantasies over the last few pages. We have entered an alternative world, all thanks to the power of language. We have even used language to ask questions about language and contemplate its transformation. Whatever the future holds, it will be, in part, a manifestation of the unending creative potential of language.

chapter 6

All About Time

Science has deciphered many puzzles. It has split the atom and discovered families of elementary particles. In turn, these families have dissolved into quarks and superstrings. As to the universe itself, scientists believe it began in a big bang, followed by a period of very rapid expansion. They have also uncovered clues about the origin of life and believe that it may extend throughout the universe. They have discovered encoded instructions for growth and development within the DNA molecule. Some scientists even believe they are piecing together clues about the nature of consciousness. But one great mystery remains untouched by the hand of science, and that is the mystery of time.

The physicist John Wheeler summed up the problem when he asked why the equations of physics don't get up and fly. We can write down our very best laws of physics, even the most recent speculations, yet they simply won't "fly." They don't leave the paper to express a dynamic, living universe. In other words, the essential nature of a truly dynamic time is missing from physics. If time's real nature is ever pieced together, the implications will be even more revolutionary than those of quantum theory and relativity. And within the mystery of time may also reside some clue as to the source of creativity.

Now it is certainly true that a particular notion of time is already used in physics. But this is only a pale reflection of the time we experience and in which we live. When Isaac Newton wrote his *Principia Mathematica,* he wanted to place the laws of nature on a strictly logical basis. In doing this, his model was the philosopher René Descartes. Newton wanted to do for physics what Descartes had done for the investigation of rational thought. In the preface to his book, he wrote, "I offer this work as the mathematical principles of philosophy" and "have endeavored to subject the phenomena of nature to the laws of mathematics" (*Philosophiae Naturalis Mathematica Princiopa,* 1687).

Everything in the *Principia* had to follow rationally from what had gone before. The great physicist did not wish to make hypotheses or indulge in speculation. Nothing was to be arbitrary or left unexplained in his new physics. But space and time presented something of a problem in Newton's great scheme. He realized that "common people" understand such notions through their experiences of "sensible objects." Such "prejudices" were to be avoided at all cost, for Newton wished to rise above the "relative, apparent and common notion of time" to the concept of "Absolute, true and mathematical time." Such time "of itself, and from its own nature, flows equably without relation to anything external" (*Philosophiae Naturalis Mathematica Princiopa,* 1687).

For Newton, absolute time and space became the backdrop against which the universe played itself out according to the laws he himself had discovered. But time lay outside the domain of these laws. It was untouched by them, as it was untouched by matter and energy. As the universe changed, time did not respond to this change. Time simply flowed "of itself," without any dynamic, interactive engagement with the rest of the universe. In short, time was fated to become a mere parameter or an algebraic symbol in the new calculus of Newton.

When an apple falls, it keeps gaining speed until it hits the ground. Its speed is constantly changing in time. Newton demonstrated that the *amount* by which this speed increases is the same from moment to moment. So although the falling apple's speed is increasing, the rate of this increase, the apple's acceleration, is constant.

Newton's calculus (which he called "his methods of flux-
ions") was specially designed to deal with problems like this.
The very idea of mathematical differentiation is essentially an
expression of rates of change. In Newton's hands, the calculus
was used as a tool of physics to calculate the way things move
and how their speed changes under various forces. But the
method itself proved so enormously useful that it was soon
employed for calculations using other sorts of rates of change,
from the speed at which a hot body cools to the rates of chemical reactions.

Given the speed and position of a cannonball at one instant
of time, t_1, Newton's laws and his calculus allow us to work
out its position at some other time, t_2, in the future or, if you
like, at some time, t_0, in the past. What's more, these changes
(the concentrations of chemicals, the changing speed of a
comet, the temperature of a cup of coffee) can all be displayed
on a graph. Along one of the axes of the graph will be the symbol t, measuring off time in seconds, hours, or years.

This t, this parameter of the calculus, this axis on a graph, is
of enormous practical use in physics, chemistry, biology, engineering, and a host of other fields. Before long, it became the
essence of time as it appears in science. But to what extent
does the parameter t have anything to do with the experiential
time of memory? How is it connected to creativity, growth,
and the generation of the new?

Even to ask such a question implies that we already have
some vague notion about time and how it should be represented. But time remains as much a mystery for philosophers
as it does for physicists. Saint Augustine said, "What then is
time? If no one asks me, I know; but if I wish to explain it to
he who asks, I know not." Add to this the adage that time does
not exist: The past is gone, the future is not yet, and the present is an infinitesimal moment that is already passed. So time
is nowhere!

Maybe the mystics could give us a clue to this mystery, since
they claim to have had a direct access to the inner nature of
reality. But mystics tend to be more concerned with timeless-
ness than with time itself. And what should we do with the tra-

ditional wisdom of other cultures? In ancient China, a distinction was made between our everyday world of time and "the Eternal." Works such as the *I Ching* convey an intimation of the eternal world that lies beyond appearance and material manifestation. Similarly, the world's great philosophical systems are concerned with an ideal world beyond the immediacy of the flesh and the grip of time. For Plato, all that exists is a mere shadow of the ideas or forms found in an eternal domain beyond the vagaries of time. And according to the perennial philosophy of the world's religions, time corrupts. Time rusts. Time ages and brings death. Why bother with the inner mysteries of time if all that exists in the domain of time is destined to wither away and decay?

Even within physics, there are inferences of a world that lies outside time, a domain that is more privileged and prior to the world of flux and temporality. In what region, for instance, do the laws of physics exist? They describe, or regulate, the transformations and movement of matter and energy, yet they are not themselves corporeal entities. The laws describe how things change in time but are themselves supposedly untouched by time. (This is a major assumption I will call into question later in this chapter.)

Laws of nature describe the creation of the universe in a big bang. However, these same laws must themselves lie outside time's domain. The laws of physics begin to look a little like Platonic forms. They are somehow supposed to have been lying around before time began, ready to spring into action with the big bang and then waiting to be discovered by human scientists!

The image science has created of the cosmos is curiously inanimate and frozen. The universe does not seem to move in a truly dynamic and organic way; rather, it appears merely to be a series of transformations characterized by the parameter t. This frozen nature is also present in the two pillars of twentieth-century science—general relativity and quantum theory. *In neither of them does anything ever happen!* But if this is so, how are we ever going to understand the creativity within the natural world—that force or energy that creates and renews?

At first sight, Einstein's relativity represents a breakthrough in our understanding of time. Space and time are now unified into a single space-time. What's more, space-time is not dislocated from the rest of physics, as was the case with Newtonian space and time. Space-time is modified by the presence of matter and energy. In turn, it reacts to influence their motion.

So far so good. Time, in the form of space-time, has been brought back into the universe as a participator. Time (as space-time) is not impervious to matter but responds to its presence. But what sort of universe has Einstein left us with?

An apple, stone, cannonball, rocket, or human being is described in relativity by what is called a world-line, drawn in space-time. A world-line is a sort of history. Your own world-line begins at your birth and describes, at various dates, the different places in which you lived as a child, youth, and adult. Here and there, it intersects with other world-lines that represent encounters with people and objects at specific places and particular times. In Einstein's relativity, the whole universe becomes a network of world-lines that thread their way between intersections. It is a God's-eye view of the universe in which past, present, and future are all equally displayed. Looking at relativity is a bit like looking at a road map and realizing that the top of the map is the future and the bottom is the past.

The problem with Einstein's vision is that there is nothing on this map to tell us where to locate our "now." Every place in space-time is equivalent, and there is no dynamic sense of a flowing time. There is no real movement, no progression, and no dynamism in Einstein's world. It's not even possible to say which is the past and which is the future on this map; there is nothing to say which is the top and which is the bottom.

Distinguishing past from future becomes a matter of convention. Otherwise, scientists have to import something extraneous into the theory, such as assumptions about entropy, the origin and expansion of the universe, or the thermodynamics of black holes. At first sight, relativity looked like a step forward in figuring out the mystery of time, but in the end, it takes us no further ahead.

If relativity doesn't work for these purposes, then what about quantum theory? At once, we confront physicist David Bohm's pessimistic remark, "Nothing ever happens in quantum theory" (Bohm, in conversation with the author). In essence, the quantum theory tells us about probabilities but not about the occurrence of actual, concrete events. Quantum theory expresses the potentiality for finding an electron in a particular region of space, but it can't predict exactly where the electron will be in that space at any particular time. It can, however, calculate the chance that a radium atom will decay within the next twenty-four hours; but it can't predict the actual moment that this event will take place. The realization of things, that magical moment when a propensity is turned into an actuality, lies outside the province of the theory.

So even the scientific revolutions of the twentieth century brought us no closer to the nature of time. They simply presented a frozen world or, at most, a series of inanimate snapshots. And if science is not timefull, then neither is it timeless. The inert worlds of the relativity and quantum theories are in no way like "the Eternal" of the *I Ching*. However, the consciousness of scientists who create these theories is very much part of the world of time. Surely, there must be a clue somewhere, some region of leverage in which time can enter. Maybe it lies within the "arrow of time," that mysterious force that appears to drive time in one direction, from past to future. Are the nature of this arrow and the secret of time waiting for us in the heart of physics?

The Arrow of Time

Our experience tells us that time has a direction. As soon as we buy a new car, its value begins to depreciate. After a year or two, there may even be spots of rust and the tires will need replacing. We too are aging, and no matter how much we may desire it, we can never turn back the clock and become younger. Although we can hope for things in the future and anticipate events to come, we only remember what happened to us in the past, when we were younger.

It is said that the ancient Greeks pictured the past as before them and the future as behind. This was because they didn't have eyes in the back of their heads. They could not see the future, but they were able to view the past as if it were stretching out in front of them. The future is veiled from all of us. Some believe it possible, during mystical experiences or moments of extreme stress, to break free of this constant forward movement, but it is, for most of our lives, the ambiance in which we live.

The laws of physics don't have such a built-in arrow. That parameter t in the equations contains nothing to tell it that it should be running from past to future. Hit the cue ball in a game of pool, and it makes an impact with the colored balls, scattering them across the table. All that takes place happens according to Newton's laws. Now reverse a video of the game and watch the colored balls roll inward into a closely packed formation, pushing the cue ball back up the table toward the cue. It may look crazy, but none of what happens in this time-reversed world actually contradicts Newton's equations.

The same thing applies to light. Throw a switch, and light from your lamp spreads out into the room. But the laws of light, Maxwell's equations, also allow for solutions in which light from all parts of the room rushes in and converges on the lamp at that instant when you throw the switch.

The laws of physics are all time-reversible. Yet our most immediate experience is of time as a flowing river dragging us along. Everything around us points to an arrow of time, to a time that flows uniquely from past to future. But science remains curiously silent about the underlying meaning of the arrow of time.

An arrow of time is needed in many areas of science. Biology deals in development, from the fertilized cell to the adult animal, from seed to plant. Psychotherapy is based on the idea of unlocking repressed memories—we don't have the memory first and the trauma later! Thermodynamics tells us that entropy, the degree of disorder, always increases in a closed system. But how do we know that such an increase has taken place unless it is measured against some unambiguous arrow of time? Likewise, the universe is constantly expanding and may even have had its origins in a primordial big bang. But to speak of expansion rather than contraction only makes sense if there has always been a temporal flow in one unique direction—from past to future.[1]

And so, science needs an arrow of time, a drive that keeps pushing the envelope of events from past into future. Yet science gives no clues as to what drives this forward movement, this temporal evolution. Wherever we look, time frustrates us.

But what about entropy? Thermodynamics dictates that the entropy (disorder) in a closed system must always increase. Couldn't science therefore use an "entropy clock" to indicate the arrow of time? If a system's degree of disorder has increased, then time must have been going forward.

This is a possible way out of our dilemma, and scientists have long debated the significance of entropy and disorder. But the whole issue is hardly clear-cut. One question is how to make an objective measure of disorder—one not based on human prejudices about order. But the main problem is that most physicists do not consider thermodynamics to be a truly fundamental theory. Its results are gained by averaging out over an astronomical number of small events, such as the colli-

---

[1]Time reversibility has, of course, been the subject of many science fiction stories. It has also crept into mainstream literature as a device for dealing with the extreme trauma of genocide. Many thinkers have asked how art could be possible after the Holocaust; they wonder if that event should even be treated by such a distancing device as literature, music, or painting. D. M. Thomas's *White Hotel* presents a fictional Sigmund Freud dealing with an apparent case of hysteria and recurrent nightmares. His methods do appear to reveal a deeper layer of repression, yet at the end of the book, we realize that the symptom, the neurosis, was, in fact, anticipatory of the horrific massacre at Babi Yar.

Another novel, Martin Amis's *Times Arrow,* unfolds from death to birth. It is told not so much by a narrator as by an internal witness who remembers the future and only has dim intimations of the past. The final horror is the realization that this "passenger" resides within a man who was a concentration camp doctor. Again, it is almost as if the act of writing about or acting as a "witness" of such horrors can only be redeemed by turning time around and thereby wish away the past and wipe clean the stained slate of the twentieth century.

sions of molecules in a gas. In turn, each of these microevents obeys time-reversible laws. This means that the direction of time given by entropy would not be truly fundamental but rather the result of an averaging out of events. In other words, the arrow of time would only occur on the large scale.

Some scientists, such as Ilya Prigogine, believe that the nature of time can indeed be discovered in this way. Others continue to scratch their heads about the arrow of time. Is it perhaps an emergent property, they ask, something not really present as a clear concept in the microworld but only occurring in our scale of things? Or does an expanding universe drive all other temporal processes in one unique direction, coupling together lots of miniarrows to move them in a collective way? Or could it be that some much deeper theory must first be discovered, a theory out of which quantum theory and space-time will emerge?

In the next section, I will offer one more suggestion: that there is indeed a true and absolute distinction between past and future, between a reality that is contained and enfolded within our present and one that has not yet come into existence.

Conditioning, Past and Future

Science has so far been unable to supply a deep reason for the arrow of time because it has no meaningful way of distinguishing past from future. Given the present speed and position of a comet at one instant of time, science can predict its entire trajectory—past and future. As we have seen, time is simply a parameter t in the equations and not something dynamic in its own right. The equations of physics equally tell us how the behavior of a system unfolds as this parameter t increases, from the present into the future, or as t decreases, from present into past.

However, we know through experience that there is a profound difference between past and future. The future is not yet manifest; it is still to be actualized. Although it is true that the past is no longer directly present to us, it is nevertheless enfolded within a variety of physical records that surround us. The past has been actualized, realized, and made manifest. Its shadow can be seen in everything from fossils and photographs to memory traces in the brain, writings in a diary, old love letters, eroded river valleys, and pebbles worn smooth by wind and rain.

From within our present, we can use these records to bring the past alive for us. This, in itself, is a creative act but one that is not arbitrary, for the past we re-create must always begin from and be verified by the records that the past has left us within the present.

It is also true that we can anticipate much about the future. But unlike the past, everything about the future cannot be known. We have records of the past, of course, but we do not have a yardstick for the future against which to measure our predictions. We can be fairly sure at what time tomorrow's plane for New York will leave the airport. But this prediction is always contingent upon a wide variety of occurrences. Bad weather, a strike, or some sort of mechanical fault may cause the flight to be delayed or canceled. Admittedly, science could also take such contingencies into account but only if they were contained within the overall parameters of its equations.

Let us look at the situation by making an analogy to a map. Science wants to give us a map of the world—a picture or representation of everything that is going on. This map would describe the way things move, their form and structure, and so on. When it comes to time, we can think of these maps as being bound together to form a book. If we open the book at our present, call it at page 100, we will get a map of the state of things at one instant of time, our "now." Based on this particular map, the laws of nature allow us to calculate exactly what the maps on pages 101, 102, and so on will look like; for that matter, the laws will also let us calculate the maps on page 99 or 98.

Science gives us a way of going from map to map, of discovering how the map on one page is transformed into the map on any other page. As we go from map to map, from one instant of time to another, the features change, and the maps themselves are constantly transforming. Tiny details on one map are displaced in the next one but always in such a way that a point on one map moves to some nearby point on another map. In fact, scientists use the word *mapping* to describe what is going on. They say that the state of a system (the way a system looks and behaves) at time t_1 is *mapped onto* the state at time t_2 or, for that matter, t_0.

Scientists call these "1-to-1" or "unitary" mappings. In a unitary mapping, a system changes in such a way that each point, or feature, on the map shifts a little bit—but no totally new points, no completely novel features, are added. All that is happening is that features in the maps are moving around and transforming one into the other. The maps don't get any larger, and their scale doesn't change.

Put another way, in a unitary mapping, the past is entirely contained, in an implicit way, within the present. Likewise, the future is contained, implicitly, within the present. Time is a parameter describing the way features in these maps transform and shift. In its deepest sense, there is no true creativity within such a concept of time. Nothing truly new can ever be born. What already exists is simply rearranged from map to map.

Now we can open the door to a deeper sense of time. We can go beyond the unitary mapping, the 1-to-1 transformations of present into future. The deeper nature of time can be found in nonunitary mappings. In such mappings, although the past may be totally contained (in an implicit, enfolded way) within the present, the future never can be. In going from the map of the present into the map of the future, there is always the possibility that new points will be added and new features will appear that have never been seen before. Thus, even though the parameter t can take us from page 100 of our map book to page 99 or page 50 or even page 1, a nonunitary mapping cannot take us to the fullness of page 101. Rather, it can only present a partial and incomplete picture of the map that appears on page 101.

It is true that much of the future can be anticipated from the present and that the map on page 101 is not all that different from the map on page 100, but the full potentiality of the future can never be totally contained and enveloped within the present. The future is both qualitatively and quantitatively more than a simple transformation or rearrangement of what is already known. The future contains the possibility of novelty, of a different kind of freshness derived from a mere rearrangement of the present.

Here, I have to add a word of caution, for Wittgenstein must be turning in his grave because I'm making serious category mistakes. By invoking nonunitary mappings, I am attempting to point to the sense of newness and freshness that is present in the future. But much of this is of a qualitative, rather than a quantitative, nature. I'm referring to the particular qualities inherent in time. The qualities of newness include both the radically different and unexpected as well as those acts of renewal, such as the baking of loaves in my neighborhood bakery. In this, a subjective element enters that takes us beyond the domain of science to represent. We are anticipating Wolfgang Pauli and his insistence that physics has to acknowledge or come to terms with the subjective elements of the cosmos, and among these, I include the notion of quality.[2]

To go beyond the linear time of physics requires the addition of a new ingredient—creativity. Within creativity, there is an absolute distinction between past and future: The one cannot be contained or encompassed within the other. Time assumes a new, dynamic nature as creativity drives the forward movement of time from the present into a future. And this future is always, in part, unknown. The essence of time's movement into the future lies in creativity and maybe even in what Pauli called "the irrational."

The Manifest and the Manifesting

Yet another clue to this nonunitary, creative nature of time comes from the language and worldview of the Hopi.[3] Like

---

[2]Within other traditions, quality also has an objective nature and is far from a purely personal value.

[3]I have not had the privilege of talking to Hopi elders. Neither do I have any firsthand knowledge of their language. What I express here has been filtered

some other Native American groups, the Hopi do not make a distinction between space and time but experience a unified space-time closer to that of Einstein's relativity. That means that time is not thought of in an abstract way but as the time experienced in a particular space or location.

But what sort of a time (or rather, space-time) do the Hopi experience? According to the linguist Benjamin Lee Whorf, who spent time with the Hopi, they do not live in terms of past and future, duration, or flow. Instead, their world is divided in a way not unlike that which I propose in this chapter.

On the one hand, there is the world of the manifest. The manifest is that which has been brought into existence within actuality. It is all that is directly experienced though the senses of the body. Within this manifest, there is no division between past and present.

On the other hand, there is the other world that is manifesting. It is not available to the senses of the body. It includes all that has not yet come into manifestation—what we would term "the future"—as well as what we would probably refer to as being "mental" or "imagined" or felt with the heart rather than the hands. The manifesting includes what is held not only in the hearts of human beings but also in the hearts of animals, plants, trees, and rocks.

The dynamic nature of time within the Hopi world appears to be a movement into the future brought about as the manifesting enters the realm of the manifest. In this sense, the manifesting is always greater. It is always potential. However, the manifest consists of all that has come to pass. It is the thought brought into action and realized in the world; it is the promise of rain that has been actualized as a watering of the earth. It is all that now exists in the world of the material and is equally real in the present-past. As with nonunitary mappings, the Hopi manifesting is the driver of time, a driver that also promises the unity of mind and matter.

---

through the writings of the linguist Benjamin Lee Whorf and discussions with my friend, the linguist Dan Moonhawk Alford.

Collapse of the Wave Function

Is there any way that science can contribute to this notion of manifesting and the operation of the creative in time? Up to now, physics has provided no real indications as to the nature of time. But Pauli did leave us one clue when he spoke of the need for physics to come to terms with the irrational in nature. I don't pretend to fully understand what Pauli meant by the irrational, but I can take a guess. Actually, I don't think that what he left us is a full or even a partial answer to the mystery of time, but it does make us stop and think.

Pauli's conjecture arises out of the profound difference between descriptions of the physical world in Newtonian and quantum physics. According to Newton, if you know the initial speed and position of a cannonball and take wind velocity into account, you can predict the precise point where the ball will hit the ground. Indeed, an instant after the cannon has been fired, there is absolutely no need to look at the ball in flight. Science has utter confidence about the precise point where the ball will land. And what's more, both science and common sense agree that even when we're not looking at the cannonball, it must still be somewhere, flying through space until it hits the ground. Only philosophers and the insane doubt the objective existence of the world or question if things cease to exist when they close their eyes.

Common sense flies out of the window in the quantum domain. Take the quantum situation analogous to the cannonball. An electron is shot toward a target by an electron gun (the sort of thing you find inside a television set). Why not measure the initial speed and position of the electron and thereby work out exactly where it will hit the target? Quantum theory says this can't be done. Common sense and the quantum world part company every step of the way.

In Newton's physics, a cannonball has a definite position in space and a well-defined speed even when we're not observing it. A set of numbers defines the position and the speed of the cannonball. In a sense, these numbers not only describe the cannonball's flight but also tell us something about the actual existence of the cannonball—that it *possesses* a particular speed and position. But when it comes to an electron, we can't even say that it "has" a position or a trajectory: Position and speed are not *possessed* by an electron.

The deeper we probe into the meaning of quantum theory, the more we find that it does not even make sense to ask such questions as "Where is the electron?" At one moment, the electron leaves the gun; at another, it hits the screen. But we can't really speak of it as being "somewhere" or even of its having a path in space between these two events. The very state of being of an electron is profoundly different from that of a cannonball.

So how do we describe or even talk about an electron? Quantum physicists must make do with what is called a "wave function." The wave function tells us the probability of finding an electron in each particular region of space. But it can never tell us where the electron will actually be found. As Heisenberg put it, the wave function deals in potentialities but not in actuality.

Thus, after the electron has left the gun, we can't say for certain where it is; in fact, such a statement would not even make sense. The best we can do is to give the probability of its being found within different regions in the whole of space. However, when an electron hits a screen, such as a photographic plate, it leaves a tiny mark. At the moment of impact, the electron is no longer everywhere in space simultaneously. Rather, it really *is* somewhere—it has hit a point on the screen.

Up to that instant, the electron is described by a wave function, in terms of probabilities all over space. The moment it hits the screen and is registered, it occupies a well-defined point in space. This means that the description of the electron changes in a radical and discontinuous way. Physicists call this "the collapse of the wave function," and here, according to Pauli, is where the irrationality of the universe enters.

We would hope that quantum theory gives us a complete description of everything that happens in the quantum world. It is therefore natural to ask quantum theory, as Einstein asked Bohr in his famous debates on the philosophy of quantum theory, to describe exactly how the wave collapses. In what way does a description in terms of probabilities spread out all over space suddenly become a description of an electron localized at one point? Quantum theory remains silent on this question. It refuses to answer. The collapse of the wave function lies totally outside the theory. It cannot be accounted for.

So is the theory flawed and incomplete? It is not, according to Neils Bohr. Bohr was able to answer every objection that Einstein put to him during the early years of quantum theory. Quantum theory, Bohr concluded, is complete. Nothing needs to be added to it, and it contains no loopholes. So when we ask about the collapse of the wave function, we realize we have reached a blank wall at the heart of physics. For Bohr, this is where the limits of language lie. When we ask such questions, we are trying to force ideas about the large-scale world into a domain where they do not belong. It is not that the theory is incomplete. Instead, it is that quantum nature has no room to import classical ideas from the large-scale world. According to Pauli, this blank wall is the irrational face of nature.

Pauli deserves to be taken seriously. He was one of the most outstanding physicists of his day. Along with Bohr, Heisenberg, and Schrödinger, Pauli was one of the creators of modern quantum theory, and the work for which he received the Nobel Prize was completed before he was thirty years old.

After his thirtieth birthday, he began to develop a passionate interest in the depth psychology of Carl Jung. Although he continued to be active in physics until his death, he did reveal to his closest friends and colleagues that he was involved in even more important work. As he remarked to H.G.B. Casimir, his assistant, "There must come something else. I think I know what is coming, I know it exactly. But I don't tell the others. They may think that I am mad. So I am rather doing five dimensional theory of relativity although I don't really believe in it. But I know what is coming. Perhaps I will tell you some other time" (quoted in H.B.G. Casimir, *Haphazard Reality*, 1983).

Pauli understood the significance of Jung's concept of the collective unconscious—that part of the mind that lies beyond the purely personal and belongs to something universal, to all human beings, and even to life and the cosmos itself. Jung, Pauli argued, had discovered an objective level to mind—that which lies beyond the subjective. In turn, physics had to discover the subjective level to matter—that part of nature that lies beyond the purely objective. Pauli's insights bring us very close to the notion of a creative universe. But in his scenario, the living universe is not portrayed by a poet, artist, or mystic but by a physicist.

For Pauli, the collapse of the wave function lies outside the domain of objective physics. It cannot be encompassed within the rational. It is a definite and discontinuous happening, an edge between all that is describable by means of a wave function and that moment when the future becomes the present. In this sense, it is a nonunitary moment, a creative event that lies outside the conditioned. It is the creative, something very close to the Hopi's concept of manifesting.

Pauli's notion of the irrationality of nature means that the quantum universe is far from a machine. Maybe it exercises an element of choice; maybe there is something mindlike about it. In any event, the singular nature of collapse, as envisioned by Pauli, does make for an absolute, nonunitary distinction between past and future. And, as with the Hopi's manifesting, that singular nature opens the possibility for a deeper unity of mind and matter.[4]

Let us return for a moment to the idea of nonunitary mappings. This concept involved the notion that the future holds the new, both in terms of quality and form. Although the col-

---

[4]There is an interesting sideline to this possible link between mind and time. For several years, David Bohm and Basil Hiley had been attempting to discover a deeper theory out of which could be derived quantum theory and a dynamic space-time. Their researches took them to algebra that had been developed in the nineteenth century by the mathematician Herman Grassman. Grassman's algebra, they believed, offered the possibility of deriving some sort of dynamic description of time. Later, when the physicists went back to read Grassman's original notebooks, they discovered that his approach had, in fact, been developed as an algebra for describing the movement of thought! His approach to the inner transformations of mind thus turned out to have a curious but a deep connection with the movement of time.

lapse of the wave function may well be the response of physics to this idea of an absolute distinction between past and future, there is also a sense that there is a qualitative change of value that cannot be reduced to any objective, scientific representation. Here, the irrational could again be said to be involved. Changes in value, the movement from present into future, cannot be reduced or fully imported into any abstract, mathematical order or ratio. The changes are richer and lie outside the domain of what can be expressed in number, measure, and ratio alone.

But a word of caution must be added. Many physicists would not agree with Pauli's position, contending that it is too extreme and mystical. Their own attempts to explain away the collapse of the wave function, however, stretch credibility even further.

Change and Constancy

The inner nature of time, its essential movement from present into future, appears to lie within the creative—the manifesting, the nonunitary, and the irrational. But its dynamism also requires that creativity restrain itself. If creativity were total and radical, the future would be filled with absolute novelty and be quite unlike the present. In such a situation, we would not be able to comprehend the passage of time or even the appearance of the new. All about us would be endless flux and a buzzing confusion; there would be no handhold for consciousness to grasp. Even our own memories would not make sense, for there would be no element of constancy within all this change against which to place and measure them. And as to the self, that sense of an "I," it too would vanish in the flux.

We sense the passage of time and understand the operation of creativity only when change can be measured against what does not change. In practical terms, time is measured by counting repetition against change—the swing of a pendulum, the sequence of day and night, the vibrations of the crystal inside a watch.

The creative must operate through time not only to bring about change and novelty (in such a way that the future can never be totally encompassed within the present) but also to preserve form and regularity. This is yet another example of the operation of creativity as renewal. If all that existed were total flux, with no constancy, then there would be no galaxies, no stars, no stable atoms, nothing but a Hericaletian fire dissolving all form into endless chaos. Instead, we have a universe that is ordered and stable and in which the flux of change occurs against a background of consistency. Again, Apollo and Dionysus are weaving their web within the world— the one preserving order and form, the other welcoming the irrational and radical change.

Newton's laws are an expression of this creative renewal. To speak of inertia, for example, is another way of saying that movement always takes place in such a way that it clings to form. Inertia is an expression of nature's habit contained within the flux of change. A body in empty space continues to do what it is already doing. If it is at rest, it remains at rest. If it is moving, then it continues to move with the same velocity. Add a force, such as gravity, and a body speeds up. Yet within this change also lies stability, for the amount of this increase (its acceleration) is constant from moment to moment.

Even Einstein's relativity confirms this sense of habit within nature. Admittedly, the appearance of phenomena is different for different observers. But underlying these diverse phenomena are laws of nature, and, as Einstein showed, the *form* of these laws remains unchanged (that is, invariant) from observer to observer. No matter how we move, what part of the universe we live in, or what mathematics we use, we will discover laws that have an identical underlying form.

Some scientists, such as Rupert Sheldrake and Ervin Laszlo, have speculated on an underlying mechanism that produces the constancy of these laws of habit. Laszlo invokes some sort of field of information at the quantum level; Sheldrake postulates "morphic fields" that are gradually built up in time out of nature's habits and that influence future form and behavior.

In this book, I will be content to invoke a general principle: that within the nonunitary mappings of past to future, there is always a strong element of "clinging to form." Creativity expresses itself in both the new and the renewal of what already exists. It is creativity that moves time, working within the present to produce regularity within change and, at the same time, an openness to novelty and that which lies beyond the conditioned.

Time Past

Creativity, the unmanifest, operates within the domain of time to produce the future. But a time that is creative and nonunitary also opens the door to notions of evolution and qualitative change from past to future. Earth's fossil record shows the way that species have appeared and transformed. Some thinkers have applied similar evolutionary ideas to consciousness, society, philosophy, and religion. Along with this runs an evolutionary ethos that sees human consciousness as somehow increasing and "getting better," as if it and society were moving toward a goal or end point.

The danger in all this is that we may view evolution—whether of species, consciousness, or society—in a simplistic and linear way through the blinders of linear time. The whole subject can too easily end up seeming like one of those charts that winds its way along the schoolroom wall, with dinosaurs, humanoids, Greeks, Romans, the Renaissance, the American Constitution, and the first Moon landing displayed in section after section. Inevitably, such an approach carries with it value judgments that reflect the prejudices of a society and indeed of the human species itself as some sort of crown of cosmic evolution. Though the creativity that lies within time is a theme of this chapter, we first need to free ourselves from the notion of the past as a line stretching back in a linear time.

Some have argued that the Western notion of linear time originated in the importance placed upon the Crucifixion by the early Christian church. For most societies, time has a cyclical nature, but the Crucifixion and the incarnation of Christ are historically unique events. For the Christian church, they became both an end point—the culmination of Hebrew prophecies—and the starting point for redemption. The Crucifixion is a unique marker in time from which dates are calculated and laid out in an "earlier than" or "later than" format. Add to this the appearance of the first mechanical clocks, erected on town buildings in the thirteenth century, and time becomes something measurable, linearized, and reduced to number. The notion of time as quantity has overshadowed the deeper, felt qualities of time.

Much of the world does not envision time in that way. We have already discussed the Hopi concept of the manifest, a co-existent past-present containing all that is accessible to the senses. Time is also seen in terms of great cycles and of cycles within cycles within cycles, as with the Mayan and Aztec calendars. For other societies, the past is a sacred repository of compacts with the energies and powers of the universe, significant events, and the creation of the world.

Such societies, of course, understand a linear sort of time when it involves the immediate past—a meeting this morning, rain last week. But if they consider events a little further back, things begin to merge into an eternal, copresent "past." I have tried to gain some faint sense of what this past could be while talking to Native American elders. To the extent that I could escape from my own mental and linguistic conditioning, it seemed to me that a sort of distant past coexists with more recent events. Some things may have happened over a thousand years ago—events that are celebrated in songs, stories, and ceremonies. Other events may have occurred more recently. Yet all are copresent on an equal footing.

It may be possible for anthropologists to date the appearance of the pipe to the time of the Plains Indians or the creation of the Iroquois confederation and its Great Law, but the Western sense of historical time has a profoundly different quality from that which speaks of the appearance of Buffalo Calf Woman to the Lakota or the arrival of the Peacemaker in his canoe among the Haudoshone.

In this indigenous time, all events are enfolded together within a unity called the past. Only when we attempt, within our present, to unroll the past into a series of fragmented events can anything like a linear time be comprehended. The linear time of physics now becomes a secondary view of time. They are events unfolded into the manifesting present out of the great coexisting eternity of the sacred past.

Australian aborigines speak of "the Dreaming" and "Dream Time." We in the West tend to interpret this as a mythic time in which the creators of the landscape walked across Australia. Following this time of creation, the ancestors became songs, or rocks, or features of the landscape. But our notions of "past," "before," and "after" do not capture the essence of the Dreaming. Some aborigines have told me that the Dreaming is not exclusively in the past. It is not something that has gone and been lost forever. Rather, Dream Time coexists with the present. The sacred past is still active with present time, and so, at night, the songs of the Dreaming can still be heard.

This sense of a copresent time out of which emerges a movement toward the future, an edge of manifesting, is bringing us closer to the heart of time. In this sense, time is a quality rather than a mere sequential order. Is this the womb of time out of which emerges the time of physics and the time of history, that time that can be drawn out on a line and measured in terms of duration and succession? This linear time is the time of commerce, of prediction, of control. It is the time of the city and the body politic, the time of physics and electronic communication. It is also that time of stress and anxiety to which we all feel tied.

Mozart spoke of compositions appearing to him whole. It is as if, for him, music existed all of a piece outside linear, sequential time. Yet his music required time to write it down as well as time to play it in a linear sequence as the music unfolds. Maybe time is like a story, an ancient myth held within a society but requiring and generating its own time as it is told. Maybe such a story can be told, or unfolded, in a number of different ways.

Laws of Nature

From the perspective of a new vision of time, we ask again: In what sense does the universe evolve? Is our universe an unfolding out of copresent time into a linear, unfolding time of the manifest world? Are the laws of nature eternal and untouched by time? Or is it possible that these laws and even the structure of space-time itself may have evolved in time? Up to now, physics has dealt exclusively with the notion that the laws of nature are given and eternal, being quite independent of the evolution of the cosmos. We question this by suggesting a nonunitary cosmos, a universe that always exists at the edge of a *creative manifesting.* Within such a universe, nothing is totally fixed once and forever. Everything depends upon context and is, in this sense, relative and open to ever wider processes.

Suppose the world begins within a pure, unconditioned act of creativity. The universe would be characterized by primal chaos, without any discernible order or regularity. But we have already seen that whenever Dionysus appears, Apollo cannot be far behind. Creativity restrains itself; it seeks both renewal and the new. Within creativity is a tendency to cling to form, and in this way, a regular, ordered, sequential time begins to emerge out of pure, conditioned process. And within this new order, the forms and habits of nature appear and persist.

At each moment, nature stands at a bifurcation point of entering the manifest. But once manifestation—realization within the world of space, time, and matter—has taken place, new forms and structures persist and repeat themselves. In this way, the laws of nature are born. They are expressions of the regularity that occurs within the creativity of time, expressions of the ways in which creativity seeks to preserve form. Thus, the laws of nature would not be eternal but would emerge out of time and evolve until they become fixed within their contexts.

Consciousness

Finally, we come to the question of time and consciousness. What is their relationship? Is consciousness the child of time?

Or is time the offspring of consciousness? By this I mean, Did consciousness come into existence, emerge, and evolve within the movement of time? Or does consciousness lie beyond time, and is time therefore a category that is projected onto the world by human consciousness?

Most biologists, who take for granted a linear view of time, hold to an evolutionary approach to consciousness. In their view, consciousness is an emergent property. This means it did not previously exist in earlier stages of evolution but emerged, as a radically new category, within the movement of time. Just as life and species evolved through a series of chance processes, so too did consciousness emerge.

The earliest life made a leap forward when it was able to protect, though the evolution of a semipermeable membrane, its internal chemistry from the contingencies of the environment. Next, multicellular organisms evolved, and then, in turn, certain cells in a colony became specialized to carry out specific functions. At this stage, although the organism retained a degree of openness to its environment, it also developed an internal homeostasis in order to preserve its integrity.

Through evolution in time, organisms became increasingly complex. At a certain level, they developed purposeful movement in their hunt for food and their efforts to mate and to escape predators. Internal regulation became more elaborate, and finally, the organism developed the capability to model its environment, anticipate, make plans, and even withhold immediate gratification in order to satisfy its longer-term goals.

Organisms became more flexible and subtle in their responses to the environment and in their ability to sense and read the world around them. In turn, sensors and nervous systems increased in their complexity. Information processing that was originally distributed all over the organism now became concentrated in nodes and finally a central brain that, in its turn, became highly specialized into various regions. Organisms with mobility could now engage in purposeful actions. But doing so requires complex internal models or maps of the world, models based on memories, learning, and the ability to recognize patterns of behavior. Consequently, they began to plan and anticipate and even to postpone immediate gratification to have greater long-term success within their particular environments.

A further stage of sophistication involved the ability to reflect on these models to the point at which the higher primates began to develop an internal representation not only of the environment but also of themselves. The organism could now project an image of itself and its behavior into the world and reflect upon it. It could anticipate the future and plan how to bring it into actuality. And finally, the organism came to understand the way its projected image and the models themselves are different from the act of projection. In other words, the organism reached a stage of reflective self-awareness.

This was quite a leap forward. It involved holding within the mind the notion of the self as an autonomous entity, something that acts with volition in the world. It was an identity that was, at one and the same time, both internal and external. For the concept of the self must also carry with it the notion of the self as it would appear to others. This requires the acknowledgment of other autonomous beings and of the existence of other minds, as well as an understanding of how they will react to one's own actions.

To have the notion of a self, to act as well as to be acted upon, does not appear to have been achieved by any creatures other than the higher primates. Only chimps are capable of reacting to their own reflections in a glass and recognizing the images as themselves. Like the awareness of the self, the mirror image is both oneself and something other, something external to one's own body. This sense that one is both actor and participator in the world is the dawning of awareness, self-reflection, self-knowledge, and what is popularly termed "consciousness."

For most biologists, this consciousness is the culmination of a series of emergent properties, an aspect of the ever increasing complexity of the human nervous system. Adherents of one train of thought believe this process is still continuing. Evolution, they say, has not ended but is continuing within the human race. It is not as visible as the transformations of animal species because the human body is pretty well adapted at this point. But, they say, evolution continues at the level of human consciousness—it is even present at the collective level as the progressive transformation of human societies.

The notion of evolution within human society is a keystone of Marxist philosophy. Most traditional societies see history as a cycle, but Giambattista Vico introduced the idea of evolution within the arts and sciences and the notion of a cycle of growth and decay within societies and civilizations. Hegel took things further with his depiction of evolution as the operation of Spirit through a dialectical process. His proposal was that this progress had reached its culmination with the manifestation of Spirit in its highest form—the nineteenth-century Prussian state. Karl Marx, for his part, claimed to have put Hegel on his feet by envisioning evolution in materialistic terms. For Marx, the end point had not been reached but required a further stage of revolution.

These themes of a move toward ever greater perfection or a teleology for consciousness and human society are also present in the psychotherapeutic movement. Freud's program, which aimed to bring repressed material to the surface, was about increasing the content of conscious awareness. This aspect was made more specific by Jung. Although Jung extended the depths of the unconscious below the personal and into the collective, he also pictured conscious awareness as extending, with more and more of the unconscious and collective unconscious becoming directly accessible to the conscious mind. Since much of Jung's program was directed toward the collective, it is clear that his notion of the extension of consciousness did not just apply to the lucky few who were able to afford Jungian analysis but also to society and the human race in general.

The paleontologist Pierre Theilhard de Chardin also pictured evolution in terms of the development and extension of consciousness. Just as life had once emerged from the sea to crawl, walk, and fly on dry land, so too would life eventually emerge into the noosphere—a domain of pure consciousness. Theilhard's vision was teleological, with the noosphere being a sort of end point reached when the universe becomes conscious of itself. At that point, all matter would become self-reflective and able to represent itself in terms of symbol, language, art, and music. Within the noosphere, the entire planet would become self-aware.

Such visions persist in contemporary culture and counter-culture. The most recent expression, via people such as Terence McKenna and Timothy Leary as well as in the novels of William Gibson, is of consciousness expanding through a combination of electronic technology and psychoactive drugs. The ingestion of psychoactive substances by our remote ancestors, McKenna argues, led to an explosion of consciousness. A similar explosion can occur today as consciousness begins to disembody, integrate itself with silicon technology, and inhabit ever more powerful virtual worlds. Such a consciousness will become truly global and collective as it enters the hypothetical "matrix" as sort of a fifth-generation marriage of the human mind and an AI Internet. In this way, consciousness would not only span the earth but also reach out into space to contact and merge with other consciousnesses in a sort of galactic noosphere.

Such visions are exhilarating. My caution is that they also contain, enfolded within them, the ethos of our current way of thinking about reality, the world, and ourselves. Just as Orwell's *1984* was a projection of postwar 1948 Britain, so too do these visions contain assumptions about the nature of progress and linear time and a Cartesian consciousness split off from the physical body. It is, to a certain degree, part of a tradition that goes back to the Middle Ages and rejects the world, matter, physicality, and the body in favor of some disembodied, spiritual existence. It also tends to overemphasize thought and abstraction over feelings, intuitions, and the wisdom of the body. In the following chapter, I shall argue that some of our greatest creativity is exercised outside the light of awareness and within the body itself.

But with this caution in mind, let us return to the issue of evolution in time and ask, Did consciousness emerge into the cosmos? Was it not implicit within the original scheme of things but brought about by a series of fortuitous evolutionary accidents here on Earth? Was there a period when the universe was without mind? Is consciousness another marker on the chart of linear time—like the arrowhead, plow, printing press, steam engine, and computer? Or was it always present, enfolded and emerging from the manifesting into the manifest world of linear temporality?

The ancient doctrine of *unus mundus* suggests that there is a single world combining matter and mind without division or, rather, that there is a unus mundus and that we, out of our own particular consciousness, have projected onto this world the separate categories of matter and mind. Just as a bird changes its plumage from spring to winter yet remains the same bird, matter and mind are aspects of a single world. To use another metaphor, consider the famous figure-ground paradox of the face and vase. Viewed in one way, two faces confront each other, but if we flip our mental switch, we see a vase. Both images can be present to our vision, but we only seem able to see one of the two representations at any given time. Likewise with matter and mind, we never seem to be able to embrace both aspects of the unus mundus simultaneously.

Carl Jung, in discussing the essential unity of matter and psyche, invented the term *psychoid* for that region that lies beyond our present categorizing. The psychoid is neither exclusively matter nor mind; rather, it is both. Just as matter-energy has been present from the origin of time but has expressed itself in different forms through the dance of Dionysus and Apollo, so too was what could perhaps be called protomind always present, as a unity to matter. It has also expressed itself in different ways within the manifest realm of space and time.

Contemporary physicists have been playing with similar ideas. The physicist Roger Penrose and the neuroscientist Stuart Hameroff have placed the origin of consciousness not at some point in time but as a process that occurs right at the quantum level. For Penrose, the collapse of the wave function is connected with the appearance of an extended space-time structure. It is also deeply tied to consciousness. According to

Penrose and Hamerof, consciousness can emerge when quantum processes in the brain reach a certain space-time scale.

A particular aspect of their approach is that space-time and consciousness both emerge, at the quantum level, out of some sort of eternal, Platonic realm. Thus, the structures created by consciousness—everything from archetypes and scientific theories to works of art—have a deep connection to the structuring principles of nature. It is no coincidence that forms and structures we find aesthetically pleasing have an exact correspondence to mathematics of a similar aesthetic quality. Both come from the same source.

At another level, attempts have been made to explain the regularities of nature—the way crystals grow, embryos develop, and instincts form—as an expression of an underlying field of information. Rupert Sheldrake has postulated morphic fields as a sort of encoding of nature's habits. Ervin Laszlo has proposed a field of memory and information that operates at the quantum level.

Yet another hypothesis comes from David Bohm, who, in the 1950s, had postulated an alternative to conventional quantum theory, in terms of what he then called "hidden variables." Later, Bohm came to view his approach as no more than a stagecoach stop on the road to a deeper theory. But he did extract from it a very interesting idea.

Bohm assumed that the motion of an electron occurs in response to a complex field of information. The essential feature of quantum theory is the context-dependent nature of the answers it gives. (Ask a question one way and we have a wave; ask it in another and we have a particle.) In addition, there is the question of chance and lack of predictability.

All this, Bohm argued, can be explained if we assume that information about the particular experimental setup, along with the rest of the universe, is available to the electron. The electron is able to "read" this information and respond. The novel and paradoxical features of quantum theory can be explained in this way.

This approach introduces two totally novel elements into physics. One is that the electron is not a tiny, featureless element but an entity sufficiently complex to be able to "read" and respond to information. (In this respect, Bohm used to joke that the electron must be at least as complicated as a television set!) The other is that this "information" is not passive, like information printed in a book or on a computer hard drive. Rather, it has its own activity. Thus, Bohm termed it "active information." When a measurement is made, this activity plays a role in determining how the electron behaves. When the measurement is over, some of this active information becomes inactive.

At this stage, there is a sort of Cartesian split between the electron and the information it responds to. But in a deeper statement of the theory, made by Bohm and his colleague Basil Hiley, the electron and the information are unified. Now the electron is no longer an object moving through space but a process constantly coming into and going out of manifestation. At present, the theory is incomplete, but if it is ever developed further, then not only elementary particles but also space and time will emerge out of this deeper level.

Currently, the foregoing are all quite speculative ideas. They do suggest, however, the continuing operation of creativity within the envelope between the manifesting and the manifest. In a certain sense, every quantum measurement represents a creative act in time. In turn, our ability to think of such things and to represent them in the virtual worlds of language and mathematics represents a similar creative manifestation of mind within the domain of human consciousness. In this way, mind and consciousness do not become something apart or, in their essence, different from the world in which we live. They are aspects of the one unus mundus. In the chapter that follows, we shall explore the way in which mind and consciousness are deeply integrated into the body itself.

chapter 7

Creativity and the Body

The previous chapter ended with speculations on the nature of consciousness. Did consciousness emerge as the material world reached a certain level of complexity? Did it only come into being after the universe had evolved in time? Or was mind, in a proto or embryonic sense, present from the origin of everything? Did mind not so much evolve in time as differentiate and express itself within the material world?

Such questions carry within them the assumption that mind, or consciousness, can be discussed in the abstract, that it is in some way distinct and separate from matter while also acting upon it. But it makes no sense to talk about mind without reference to ourselves as physical beings existing on this material planet. The argument of this book is that creativity is ubiquitous. It touches matter and mind equally and in such a way that no true division is possible between them. This chapter examines the way creativity flows through the whole of the body. I argue that knowledge, memory, intuition, thought, and intelligent behavior are not exclusive to the brain but take place throughout the body.

If such ideas appear at all radical, it is because, in our own age, we have found it so easy to separate head from heart. Increasingly, we deal in such a multitude of abstractions about the world that we become distanced from it. Our offices are climate controlled and in many other ways cushioned from the world outside. We interact with our friends via E-mail and the Web. While for most of the rest of the world survival depends on the contingencies of nature, we in the West become increasingly insulated.

The danger is that we come to believe that we have lost touch with our physicality; that we increasingly live in a virtual, abstract world of ideas, concepts, symbols, and images; that the events before us become pseudoevents and reality is confirmed not by the senses but through the reports of the media. In such an atmosphere, we can be excused for believing that reality, the grain of the world, has been replaced by simulacra, signs, and symbols.

To think this way is to fall into the grip of an illusion. This chapter argues that we are always in touch with pure physicality. We cannot ignore it. Catch a cold, get a toothache, miss your evening meal, and you rapidly realize just how much your thoughts and attitudes depend on your physical body.

Who we are is written into our bodies. Ordinary language demonstrates the way in which body positions reflect our behavior and attitude of mind. A person can be "sour-faced" or "a pain in the neck." If we are rigid in our thinking, we will be equally rigid in our bodies—"stiff-necked," "inflexible," and "unbending." If we are British, then we have a "stiff upper lip," we keep our "chins up," and we have a "ramrod" back.

Wilhelm Reich, a student of Sigmund Freud's, specifically drew attention to the way the body dispositions of his patients expressed their neuroses. One man kept his face in a fixed smile even when speaking of painful matters. Reich realized that this face had become a physical mask. The patient wore it in order to face the outer world, and its origin lay far back in the traumas of childhood. Reich used the term *character armor* to describe the way the pain of infancy is transformed into rigid muscular postures—the armor we wear against the world.

Thus, rather than neuroses and traumas being exclusively mental things, they have taken residence in the body. Hit a dog and it will cringe; traumatize a young child and for the rest of its life, it may walk with a stooped back. Just as computer data are stored as magnetic segments of a hard disk, so are traumatic memories stored with fixed postures and dispositions of the body. It was Reich's belief that these buried memories could be accessed by a form of massage that involves the vigorous manipulation of the areas of rigidity. For Reich, the way to mental health lay within the body itself. (Actually, Reich went even further than this. He believed that rigidities occurred not only at the individual level but also throughout a whole society. Thus, what may be expressed in one person as a stiff neck and an inflexible posture will, in an entire society, be expressed as fascism. The work that was to get him into conflict with authorities in Germany and later the United States could perhaps be called a holistic therapy directed to the postures of governments and entire societies.)

Another therapist, Moshe Feldenkrais, also pointed out the way rigidities in the body can correspond to similar rigidities in thought and behavior. A person whose whole attitude to life is out of kilter may not be able to sit straight in a chair, or she may have difficulty performing certain movements with ease. Feldenkrais's method was to perform minimal movements—the smaller the better—or, in fact, no movement at all while just suggesting that the person be aware of a disposition or intention to move that is suspended on this side of action. Working this way, he believed, would make it possible to achieve a corresponding freeing of body and mind.

A similar insight comes from the work of Stanislav and Paul Grof. The Grofs were pioneers in the medical use of psychedelics. When LSD ceased to be legal, they developed an alternative called "holotropic breathing." Holotropic breathing is a Western medical adaptation of methods used in certain shamanic cultures. It is analogous to a form of hyperventilation but is practiced for several hours and accompanied by loud music. This technique produces a considerable flow of energy that cycles through the body, where it encounters a variety of psychophysical blocks.

In the early stages of holotropic breathing, the subject experiences a change in consciousness but with more of a sense of volition than occurs with psychedelic drugs. He may, for example, choose to float above the experience and observe the changes in mind and body in a nonjudgmental way, observing traumatic situations from the past but without the corresponding sensations of pain and fear. However, he can also relinquish his usual sense of self and throw himself totally into the experience.

After monitoring many holotropic sessions, the Grofs became convinced that memories are stored in various regions of the body's muscles and organs and can be accessed by manipulating areas of tension during altered states of consciousness. In this, they were in accord with Reich. But then they went much further. The Grofs believed that the body contains memories within its very cells that go beyond what has been personally experienced and touch something closer to the Jungian collective. The appendix, for instance, is a biological residue (a memory, if you like) of an original second stomach—the sort used by ruminants. In a similar way, the cells of our bodies, according to the Grofs, contain encoded records of our evolutionary development and can be accessed through holotropic breathing or one of the many shamanic practices of other cultures.

If all this seems to be getting a little too far out, let us return to something more concrete—state-specific memory. In extremely traumatic situations, such as a car crash or a violent personal attack, the victim loses all memory of what took place. Classical Freudians would probably say that because of the high degree of trauma, such memories have been repressed from consciousness. There is, however, an equally plausible theory based on the idea that memories are a function of the state of the entire organism. Put another way, memories are associated with a particular state of arousal and the body chemicals that happened to be coursing though a person's body at the time of trauma.

In the fraction of a second preceding a car crash, a person will be in an unusually excited state. Adrenaline and a variety of other chemicals are flooding the body to produce a state of "fight or flight." In that exceptional body state, sense data are being recorded as memories. Later, in more relaxed surroundings, the body returns to normality, and these memories are now inaccessible. It is only by deliberately entering such a body state again—through mental imagery of the situation or via hypnosis—that access can be regained. In such states, the degree of recall can be striking: The victim can give the number of a car license plate, a highly accurate description of an attacker, and so on. The police have used such methods in an effort to solve crimes. (A degree of caution must be used here to differentiate the evocation of state-specific memories from what is known as "false memory syndrome." The latter is another example of the wonderful creativity of the human imagination. Given an overly enthusiastic therapist who is obsessed with the idea of multiple personalities or infant abuse, the patient's unconscious mind is only too willing to pick up on the subtle clues given by the therapist and invent elaborate stories of childhood abuse.)

I once met a therapist who radiated a degree of disturbance about her person. One day, she told me that she specialized in multiple personalities. When I asked if all her clients were referred cases, she quite innocently told me that they were all perfectly average patients who had come to her off the street, as it were, and that only after many sessions did their multiple personalities begin to develop; in many cases, stories of childhood abuse were also related. Of course, sexual and physical abuse does tragically occur in some families but by no means in all. Another therapist, who presented an almost messianic fervor, told me, with quite a degree of delight in his voice, of the cases he had unearthed involving murder, cannibalism, and satanic rites.

Something curious and disturbing is at work here, possibly the modern equivalent of the work of witch-finders in the Middle Ages. In many ways, this tendency is a perverted tribute to the powers of the human mind to create, invent, and project elaborate fantasies into the minds of others.

Today, a wide variety of physical therapies, based upon massage and the manipulation of the body, are designed to discharge the rigidities and pain associated with early traumas. Even voice therapists are able to detect these early injuries in the way people breathe and speak and then extend the range and flexibility of the voice through various therapies.

But if physical therapies are so useful in dealing with the pains and trauma of our earliest life, why bother with the famous "talking cure" first developed by Freud? Here again, Freud himself has undergone revision and expansion. The therapist Eugene Gendlin, for example, has written about what he calls "the felt edge." This is encountered in therapy when a patient feels that something important is hovering just on the borderline of conscious awareness. For Gendlin, it is important to bring this feeling into the light of awareness by giving it a name—by addressing it verbally. As this happens, the patient feels a strong physical sense of release and recognition. It is a somatic "aha" sensation associated with a therapeutic breakthrough. With the felt edge verbally symbolized, a new movement is possible as feelings and memories that have been trapped in the body begin to surface. The act of naming or symbolizing is a highly creative process and one that now allows for a further level of unfolding within a person's life.

The French psychiatrist Jacques Lacan has argued that many of the psychiatric symptoms he observed were literally words trapped inside the body. Symptoms, for Lacan, were created out of words. Indeed, he believed the unconscious itself is structured like a language. At its most successful, Freud's talking cure addresses this level of encoding in the body by working with words and feelings together. In some senses, this cure returns Freud to his origins and his early training with Charcot, who had demonstrated that various forms of paralysis were not the result of organic damage to the body but the body's expression of early traumas—that is, hysterical paralysis.

These examples show that our memories, conscious and unconscious, are not passive records stored in the brain but are distributed throughout the entire body. Early traumas are part of a process that has been arrested and frozen. These fossils of pain resist the creative, unconditioned movement of time entering matter. They are wounds of our creativity. But once they are released, through meditation, massage, manipulation, or verbal symbolization, the creative process is freed.

So much for psycho- and physical therapy. Only some of our embodied memories are traumatic, those that produce tensions and rigidity. The rest are joyful, productive, and informative. They go beyond memories, as an aspect of the past that has been fixed, into ongoing processes. This intelligence of the body may, at times, be superior to that which we normally associate with "thinking." After all, it is through the body, as much as the mind, that we know the world.

There is a division between East and West as to the method by which we should attain knowledge and come to understand reality. The Western view—that reality is best understood by examining the external, material world—was made explicit by the philosopher Francis Bacon, who went so far as to suggest that nature should be put on the rack and questioned!

But Western science is not the only field in which close attention is paid to nature. The Renaissance saw the birth of an approach to painting and sculpture that took, as its origins, the careful observation and recording of the external, visual world. The art of other cultures is more concerned with inner vision, symbol, decoration, and pattern. The German philosopher Rudolph Steiner, for example, suggested that during the golden age of Greek art, sculptors did not work from real human models but rather attempted to depict the body as an ideal.

In the East, wisdom is often associated with an inward journey, meditation, and a search for the truth that lies within. For the truth seeker, whether Hindu, Buddhist, or Sufi, the yardstick of truth lies in verification through inner experience. For the Western scientist, by contrast, it involves careful experimenting and measurement. Of course, this division is not hard and fast: India and Central America, for instance, have their doctors and astronomers, and the West has its mystics and inner journeyers. Nevertheless, these two worldviews can be very roughly characterized as outward looking in the West versus inward looking in the East.

But what if these seemingly different methods are no more than two sides of the same coin? David Bohm was a physicist, the practitioner of a subject born out of careful observation and experiment. Yet he also believed that understanding of the world could come about through contemplation and inner investigation. His body, Bohm reasoned, was composed of the matter of the universe and contained within it all the forms, processes, and patterns that animate the cosmos. For Bohm, the laws of nature could be discovered by looking outward equally as well as by looking inward.

As a student, Bohm noticed that ideas in physics, such as the motion of a gyroscope, corresponded to sensations and dispositions within his own body. To take one example, the way spinning electrons combine in quantum theory is not intuitive and seems to betray common sense. He described his "thinking" in the following words: "I had the feeling that internally I could participate in some movement that was the analogy of the thing you are talking about. I can't really articulate it. It had to do with a sense of tensions in the body, the fact that two tensions are in opposite directions and then suddenly feel that there was something else. The spin thing cannot be reduced to classical physics. Two feelings in the mind combine to produce something that is of a different quality. I got the feeling in my own mind of spin up, spin down; that I was spinning up and then down. Then suddenly bringing them together in the x direction . . . horizontal. . . . It's really hard to get an analogy. It's a kind of transformation that takes place. Essentially I was trying to produce in myself an analogy of that, in my state of being. In a way I'm trying to become an analogy of that—whatever that means" (quoted in David Peat, *Infinite Potential: The Life and Times of David Bohm*, 1997).

Bohm believed that his physical body contained information about the structure of the universe and its processes. But what he experienced went beyond the notion of the body as a passive container of memory and knowledge. His body was engaged in an active process whereby structures and dispositions were turning into thoughts.

A cynical, hard-nosed physicist may take Bohm's example with a grain of salt or feel it is something of an exception. But let us look a little deeper at the whole question of how science gains knowledge of the world. Einstein wrote about the way he arrived at the theory of relativity, and that process was very different from the way in which the subject is "logically" presented in physics texts. It also contradicted explanations about science advanced by the logical positivists. According to the latter, a theory is constructed in the most direct and economical way possible from a set of experimental observations.

The conventional account of relativity would begin with Michelson and Morely's experiments that showed the speed of light is constant. This led to the abandonment of the idea of an ether that permeated all space and opened the door to the theory of relativity. But Einstein did not attach too much importance to this experiment at the time. His route was quite different. His insight did not arise from an examination of experimental results or new observations but from looking into the heart of physics and seeing what was wrong.

His reading and thinking during this period included the works of philosophers Ernst Mach, Immanuel Kant, and David Hume. He reflected on the way the mind constructs its concepts of space and time. His creation of the theory of relativity was the result of a new perception, an exercise in creative imagination, rather than an attempt to match theory with experiment.

For Einstein, this imaginative thinking had a strong somatic basis. While working on his field equations, for example, he would squeeze a hard rubber ball. The muscular tensions he experienced in his arm were part of his thinking. Out of this creative act, Einstein then had to transform his insight into a theory that was logically consistent and could then be left to others to verify experimentally.

Einstein went even further in arguing that a theory is a form of creative perception, rather than the result of a passive gathering of data from the external world. He told the young

Werner Heisenberg that a good theory should tell physicists what to observe, not vice versa.

Working on the threshold of modern quantum theory, Heisenberg and his colleagues were trying to make sense of all the observations and data that were being gathered about atoms. Could this data somehow be fitted together and built into a theory? Einstein cautioned him that rather than the data suggesting the theory, the theory should suggest the data. Put another way, the theory, a product of pure imagination, should direct the physicists to the experiments they need to perform and the data they must gather. Einstein's clue led Heisenberg to formulate his quantum mechanics, in which the "observables" become patterns of numbers manipulated in special ways by the theory. It is the theory that now suggests what is observable and what is not. In the last chapter, for example, I mentioned that although position in space is a quantum mechanical observable, time is not. Time is something suggested by the theory directly and not inferred from experiments.

It is certainly true that some discoveries come about by accident—as happened when Alexander Fleming observed the way cultures were dying on a petri dish and realized the power of penicillin to kill bacteria. But even in that case, Fleming's understanding of medicine disposed him to look at things in a certain way. As Louis Pasteur put it, chance only happens to the prepared mind.

Today, elementary particle accelerators are built by international teams of scientists for enormous sums of money because theories suggest the particular energy range scientists should look in for new, exotic particles. Einstein's theory of relativity indicated that light should bend under the influence of the gravitational field of the Sun, which led astronomers to mount an expedition to observe a total eclipse of the Sun and detect this bending of light. Another factor predicted by Einstein was the slowing of clocks when they are accelerated or in a gravitational field, and as highly accurate atomic clocks became available, physicists made the relevant observations. Without a theory to suggest that such phenomena would exist, no one would ever have gone out of the way to set up experiments to observe them. (Incidentally, things can also work retroactively. Astronomers at the end of the nineteenth century had observed anomalies in the motion of the planet Mercury. A variety of ideas had been offered to explain away this effect, but it was only thanks to Einstein's relativity that this observation made sense. With relativity, the observation had somewhere to fit in; it was part of a new way of thinking about the universe and took its place alongside a number of other new observations.)

At first sight, Western science seemed to be about looking externally and gathering data from the world. But now we see that it is really about the way the inward vision teaches us to see the world in a fresh way. In turn, the subsequent metabolism of new observations and data leads to a further series of insights. Rather than considering the world as purely external and objective, the greatest scientists work in a cyclic process involving both an inner and an outer movement.

What is true of science is equally true for art. Renaissance art took as its starting point a representation of the external world. Yet even the most "naturalistic" representation must also have an inner correspondence. Second-rate painters may slavishly copy from nature, but a Rembrandt, a Turner, or a Cézanne goes much further. Such artists are concerned with a truth that cannot be exclusively restricted to the inner or outer worlds alone but embraces them both and heals the way we have fragmented outer from inner. Nature has always been the great teacher for artists, yet their relationship with nature is based upon the richness of their inner world, as well as their connection with the whole history of painting. Recall from Chapter 5 the way in which perception is an intentional act. What we see depends upon the context in which things are presented and what we already know about the world. Just as a scientific theory tells us where to look for our data rather than responding passively to the world, so too artists suggest to us a way to look.

"I have always claimed that painting's prime function is to dictate to us what the world looks like," writes the English painter Patrick Heron. "What we imagine to be the 'objective' look of everything and anything is largely a complex, a weave of textures, forms and colors which we have learned, more or less unconsciously, from painting, and have superimposed upon external reality. The *actual* 'objective' appearance of things is something that does not exist." Heron points out that what we see is a flood of impressions on the retina that must be ordered, and as he observes, "historically, it is painting that supplies that order" ("Solid Space in Cézanne," in *Modern Painters*, vol. 9, no. 1, 1996). So a change in the order of painting produces a corresponding change in the way we see the world.

We see again that the reality in which we live is presented to us through processes that are largely internal. Let us look a little closer at this perceptive intelligence of the body and the nature of its creativity.

In Chapter 4, we looked at the work of the sculptor Anish Kapoor. A few years ago, I was talking to him in his south London studio. Looking around, I asked him how his pieces were made. He described how the initial approach begins with a feeling of intention. It is a period of creative suspension, a time in which nothing is being manifested externally but a great deal of internal work is taking place. This period of holding, or containment of the intention, continues to the point at which the work is ready to be made. Kapoor described this latter stage as a stepping aside of the self—the ego-driven self with all its desires, memories, and histories.

Kapoor's work is not about the personal and idiosyncratic. Its concerns are universal, and they move beyond the exclusively Western issues of art history. Neither are they biographic pieces. They succeed to the degree to which they escape the personal and idiosyncratic. Of course, as the work is actually made in the physical sense, Kapoor must exercise choice, discrimination, aesthetic judgment, and a high level of skill. Some of his works are successful. Others remain on the studio floor for months, seeking their realization. The very best of them, the very best of all art for Kapoor, are touched by an alchemical transformation that takes them beyond anything the artist could have consciously planned or intended.

This absence of the biographical and purely personal is present in Stravinsky's account of the composition of *The Rite of Spring*: "I had only my ear to help me. I heard and I wrote down what I heard. I am the vessel though which Le Sacre passed" (quoted in Robert Craft, ed., *Dialogues and a Diary*, 1968).

The containment of a work of art within the body during a period of silent but highly creative work was also experienced by the composer Sir Michael Tippet. Tippet's major works include symphonies and operas. Because of his sympathy with the depth psychology of Carl Jung, much of his work evokes powerful archetypal structures. Among the general public, he is best known for an early work, *A Child of Our Time*, which considers inhumanity, specifically the Nazis' persecution of the Jews, in terms of the light and the shadow and the need for a resolution—the acknowledgment of the "dark brother."

Composition, for Tippet, is akin to invocation. In this, he echoes Thomas Mann's character Aschenbach, whom we met in Chapter 2. Just as Kapoor builds up and holds the tension inherent in a work's intention, so too does Tippet invite the god or goddess to be indwelling. Speaking to me about his opera *The Midsummer Marriage*—a work of mythic proportions that deals with the mystical marriage of masculine and feminine and journeys into the Dionysian and Appolonian orders—he described how he had to hold that music within him for many months.

The music and libretto for Tippet's opera are filled with such tensions that a containment of these powerful forces was necessary during the composition of this work. But during that period, no music was actually written down or made manifest in linear time; rather, it was contained within Tippet's body. In fact, it was being contained to the point at which it could then be externalized and written down as a musical score. Of course, at this point, Tippet had to use all his musical skills to release his inner processes into the world of sound. But in a very deep sense, much of the creative composition had already taken place within the body and outside the light of awareness. In the case of *The Midsummer Marriage*, the strain of holding this inner tension was so great that, at one point, the composer was suspected of having cancer. In this, Tippet identified with the considerable physical stress Berlioz experienced when composing *Les Troyens*.

Stravinsky's observations on the sheer physicality of music complement Tippet's. "What is the 'human measure' in music? . . . My 'human measure' is not only possible, but also exact. It is, first of all, absolutely physical, and it is immediate. I am made bodily ill, for example, by sounds electronically spayed for overtone removal. To me they are a castration threat" (quoted in Robert Craft, ed., *Dialogues and a Diary*, 1968).

Another composer, Alexander Goher, speaks of composition beginning with an intensely troubled period of disappointment, frustration, and inner rage. But after passing though this suffering, he says, "the music writes itself. . . . There is no longer a composer who pushes the material about, but only its servant, carrying out what the notes themselves imply. . . . For me such experience exceeds all other satisfactions that I know and can imagine. For, at this moment, I find myself overcome by an oceanic sensation of oneness with all around me" (*The Independent*, June 1, 1991).

For the philosopher and aesthetician Susan Langer, music is not self-expression but "formulation and representation of emotions, moods, mental tensions and resolutions—a 'logical picture' of sentient, responsive life, a source of insight, not a plea for sympathy" (*Philosophy in a New Key*, 1960).

The painter Patrick Heron gives a precise account of the inner feeling he experiences:

As a painter, I can testify to the following sequence of sensations: the sudden apprehension of the form of a new picture is first registered, in my own case at any rate, as a distinct feeling of hollowness: and to locate this sensation somewhere in the region of the diaphragm is not to indulge in a pretentious whim: it is merely to acknowledge physical fact. I am noting possible subjects all day long, every day, quite involuntarily. Thus it is not a question of painting when I see a subject: it is a question of calling up a subject (or to be more precise, of calling up an immense variety of remembered subjects simultaneously) when I am ready for action with my brush and palette. So I begin with this hollow feeling. Next, this uncomfortable sensation in one's middle grows into a sort of palpitation, which, in turn, seems rapidly to spread upwards and outwards until the muscles of one's right arm (if one is right-handed!) become agitated by a flow of electric energy. This energy in one's arm is the prelude to painting because it can only be released by grabbing a brush and starting to paint (quoted in A. S. Byatt, "An Act of Seeing," *in Modern Painters*, vol. 11, no. 2, 1998).

The point is that the act of creation begins outside the direct light of conscious awareness and in the body itself. One senses, feels, or has the intuition that something significant is happening and that it is vitally important that this as yet unidentified process be held and contained. The actress Glenda Jackson described her own work on building a character. For her, there was a latency period she called "putting the bread in the oven."

Heron echoes Cézanne. Speaking of how he felt when seated before nature, he said, "I am becoming more lucid before nature, but always with me the realization of my sensations is always painful. I cannot attain the intensity that is unfolded before my senses. . . . Here on the bank of the river the motifs multiply, the same subject seen from a different angle offers subject for study of the most powerful interest and so varied that I think I could occupy myself for months without changing place by turning now more to the right, now more to the left" (quoted in Richard Shiff, "Cézanne's Physicality: The Politics of Touch," in Salim Kemal and Ivan Gaskell, eds., *The Language of Art History*, 1991).

One must trust one's sixth sense during this period and not try to rush things or have them crystallized too rapidly. This is often an extremely frustrating time, particularly as the world outside places so many demands on us to produce or give progress reports.

A clue that much of our creative work takes place within the body comes from brain scans carried out on professional musicians. They produce an unexpected result, for instead of the act of listening leading to enhanced activity within the corresponding area—"primary auditory activity," as it is called—activity in that area is actually suppressed during the playing of an instrument! This suggests that the origin of music lies, at least in part, within an internal order. It is born out of a concerto of muscular tensions and predispositions to movement associated with flows of thought and emotion. Making music is the outward projection of these inner sensations through highly coordinated and subtly controlled muscular activity.

When playing an instrument, it is as important to feel what is going on inside as it is to listen to the physical sound one is making. An interesting example is given by the Canadian pianist Glen Gould. While practicing a piece of Bach, he was repeatedly frustrated in his attempt to realize the music he felt within. Finally, he placed a couple of radios on top of the piano, tuned them to different stations, turned them up to full volume, and began to practice. Up to that point, the sound he had been making on the piano was getting in the way of his attempts to coordinate the movements in his hands with the flow of his internal sensations. Now he could realize what he felt inside and begin to transform it into a musical performance. For Gould, a strict inner musical logic was coupled with intense emotion as the source of his music.

But here, a word of qualification is needed. The performer cannot rely on the body alone, for music depends upon the physicality of sound. No matter the degree to which a musician is "inspired," no matter how successfully he or she is in evoking the performance out of the physicality of the body, it is also necessary to preserve a critical and detached attitude toward what is being produced. As with an actor, the performer must learn both to give heart and soul to the music and, at the same time, remain detached from this process, watch for excess, practice fine discrimination, and judge the overall form and gesture of the performance.

Can this inward sensing also work for orchestral musicians or chamber players, who must be actively listening to their colleagues? It appears that a great deal of what is going on involves a response to each other's gestures, breathing, and so on. String quartet players who have worked together over a long period report an almost psychic empathetic connection. Paul Robertson of the Medici String Quartet noticed the way his colleagues were able to sense when one of them was about to make a mistake—even if that mistake did not subsequently take place!

This primacy of inward listening and empathic connection could explain the abilities of musicians who are profoundly deaf. The percussionist Evelyn Glennie, for one, has made an important international career without being able to "hear" the music she makes. When asked how she was able to tune members of the percussion family, Glennie replied that she "felt" each note as a sensation within particular parts of her body and was able to bring everything into harmony though a sort of inner coordination.

Therese Schroeder-Sheker originally studied the medieval harp. Her interest in medieval music led her to an approach developed in the Cluianic monastery of the tenth century. The Cluniac order opened Europe's first infirmaries. In addition to using herbs for medicinal purposes, members of the order employed music, particularly during the last hours of a person's dying. Schroeder-Sheker took this practice into a modern hospital setting, and today, she is a professor of thanatology and trains others in this approach. While monitoring hundreds of cases, she noticed that even people in deep coma responded to her music. For example, vital body signs changed when different musical modes were played.

For Schroeder-Sheker, hearing is not confined to the ears alone; instead, the entire skin is an organ of perception bathed in what she refers to as "tonal substance." Music functions within the entire body rather than being confined to a direct link from ears to brain. In this sense, the healing aspects of music go deeper than we normally suspect. Music therapy is well established and has been used in a wide variety of contexts. Generally, the assumption is that this music is "heard" though the brain, where it evokes a variety of thoughts and emotions. Music plays a role in orchestrating these evocations, extending their range and bringing them into harmony. In this

way, music exercises a therapeutic function, one that is as ancient as music itself.

But Schroeder-Sheker's work adds a new dimension to this
approach. It suggests that music is not just operating upon
feelings, thoughts, and emotions but also that it acts directly
within the body. She has speculated that tonal substance may
even work at the cellular level. Her metaphor is alchemical—
referring to the ancient practice of separating the subtle from
the gross. If this notion is indeed true, then music has a powerful effect not only in bringing thought and emotion into harmony but also in helping to heal the split we have made
between body and mind.

Projective Identification

The physical body is an arena of complex, interacting systems.
These include the immune and nervous systems, flows of hor-
mones, metabolic processes, and the various aspects of homeostasis that maintain levels of oxygen and blood sugar, right
down to the sodium-potassium balance of individual cells.
Each of these systems could be said to have an intelligence of
its own. Each is capable of recognizing complex patterns. One
example is the way the immune system recognizes what is
harmful in the environment and defends the body while simultaneously recognizing what is friendly and what belongs. All
these systems also interact with each other, along with the various muscular tensions, dispositions, and orientations of the
body.

These processes, movements, dispositions, and orientations
of the body, along with the brain, store aspects of our memories, perceptions, and feelings. New sensations are metabolized within the body, as much as they are stored within the
brain. In turn, memories and sensations give rise to yet other
processes that move through the body and mind in coordinated ways. Much of our daily life and creative work operates
from this level. The process is only dimly sensed. But in the
case of a particularly intense creativity, such as that experi-

enced by Tippet, holding the sensations for the necessary length of time can even be physically painful.

The alchemy metaphor may be useful here. Alchemy involves processes of dissolution, crystallization, and sublimation and the various stages of refinement that separate the gross from the subtle. These take place in the alembic, or alchemical, retort. The processes all need their own time, just as the body has its time for metabolization. They cannot be accelerated or bypassed. In particular, the work must take place in silence, for the alchemist must not dissipate the "great work" by discussing its nature with others.

The work of the body is similar. It must be contained and focused. The necessary heat must be generated, and one stage must be completed before another is begun. In an alchemical working, the practitioner recognizes the completion of each stage by its particular color. Similarly, the creative person begins to learn the order of working within his or her own body. And just as the medieval alchemist kept silent about "the work," every creative person knows how dangerous it is to reveal and thereby dissipate an idea too quickly.[1]

I am suggesting that the first stages of creative work are contained within the body, out of the direct light of conscious awareness. The next stage requires externalizing the work into the world of the manifest—the world of space, time, matter, and society. In the case of a dancer or an athlete, this could take the form of physical movement. For a painter, it could be the gesture made by the brush. Earlier in this chapter, we saw how Patrick Heron's work began with an uncomfortable sensation that grew into "a sort of palpitation," which spread upward and outward until the muscles of his right arm became

---

[1]In his autobiography, *You've Had Your Time*, the writer Anthony Burgess records the pleasure he found in discovering the pubs of Dublin during his late middle age. How congenial it would be, he felt, to live in such a city and drop in to the pub for a few drinks and some talk each morning. But then he realized that if he did this, he would never write again. By talking about his ideas, he would be sending them out too soon into the world, and they would lose their magical potential to become transformed into novels.

"agitated by a flow of electric energy." For others, manifestation may be spoken, written, or even voiced internally.

Let me give my own experience, which has been more or less the same whether I was doing theoretical physics, writing a play, or thinking about a new chapter for a book such as this one. First, I arrive at something akin to a question. This is not necessarily a question that could be verbalized in a precise way. Sometimes, it is a sort of pressure, or edge, or sense of suspension. When an actual verbal question occurs, then I may try to answer it, but more generally, I have a sensation, a sort of orientation, that may be similar to that instant before one moves a hand or takes a step forward.

Typically, I will hold this sensation and not try to bring it to realization. If I am faced with writing a new section in a book, for instance, I may experience several days of frustration during which I find it impossible to write or even know what I want to say. But then one night, things will begin to work. Maybe I am awakened by a thought, or perhaps when I wake in the morning, I soon find myself typing several paragraphs at high speed, trying to get the material out before I'm interrupted or distracted. Only later will I read it over and begin to edit the text. At this point, it is almost as if I am reading pages written by a different person but in my style. The words seem totally fresh and, on occasion, quite surprising, like nothing I have thought up to that point.

Projection, or external manifestation, involves the symbolization or encoding of processes that had previously existed within the body and nervous system. They now find their realization as words, symbols, gestures, structures, actions, movements, and patterns of behavior. At this point, because these processes have become external to us, we can observe and reflect on them within the conscious mind. Because they are external and symbolic, they can be mentally manipulated, played with, rearranged, and built upon. In turn, these exterior symbolic projections, these appearances in the external world, are internalized. They are ingested back into the body, where they continue to work until they can be projected outward again.

In the early stages of creation, all that may exist is an intuition, disposition, or intention that must build up in intensity while it is held internally for hours, days, or even months. Finally, with the first stages of the alchemical work completed, the work is ready to be projected into the world, externalized and symbolized. Now it can be observed, reflected upon, transformed, and integrated with other material. As the cycle continues, the work is reabsorbed, subjected to additional alchemical transformation, and projected outward again. In the heat of creation—painting, composing, writing, dancing, making a speech, and so on—this becomes a continuous cycle, with each stage overlapping the next.

This symbolic externalization, expressed as a painting, a scientific theory, a novel, a string quartet, a dance, or the like, is now public and available to society as a whole. But this does not mean that the creative process has necessarily ended. It can continue, this time being absorbed into and projected out from the body politic and its actions.

Healing

Thus far, our examples of corporealization and the sensations of the body have referred to exceptional artists, scientists, and musicians. But the creativity of the body is available to all of us. It is even enfolded within our everyday language. We speak of someone being a "royal pain." At times, we have "butterflies in the stomach" or "a sinking feeling." A broken love affair gives us "heartache." We get "fed up" when we are so "stuffed" with information that we can't "digest" new ideas. We find things "hard to swallow," and a situation can make us "sick to the stomach." We can feel "disoriented," and if things move too fast, they make us "dizzy." Some days, we "lack energy." A business proposal can "leave us cold" or even have a "funny smell about it." Other events "hit us straight between the eyes," "leave us with a bitter taste in the mouth," or "make us want to puke." A person sends "shivers up the spine" or makes our flesh "crawl" and our "hair stand on end." Sometimes, we have the "flavor of a new idea" or "a taste of freedom."

Examples such as these suggest that we're pretty sophisticated when it comes to understanding the body feelings and the sensations evoked by the events around us. In many ways, these sensations are more important, vivid, and immediate than the way we think about and analyze things. If our judgments were purely logical and rational, they would lack the underlying "gut feeling" that pushes us into positive action. However, we should not rely upon feelings exclusively. If they are not backed up with thoughtful consideration, a situation could become confused and get out of hand.

Many of the body feelings just described are not particularly comfortable. This suggests that situations that have an impact on our lives cause us a degree of bodily discomfort. But this is only to be expected. Example after example within this book suggests that creativity and change are double edged. On the one hand, there is the unconditioned movement of creativity within matter and time; on the other, there is the need to "cling to form" and preserve structures.

Resistance to unconditioned change is expressed in the inertia of traditional societies that must ratify the new within the context of the old. It is present in the homeostasis of organisms and in organizations that need to preserve their inner structures and processes intact.

Interventions from the outside are associated with sensations of discomfort because systems are required to respond and adjust in new and complex ways. This does not mean that the situations in themselves are undesirable; rather, they are acting to extend the range of our feelings and sensations. If you go skiing after the first snowfall of the season or take a long bike ride in early spring, you'll certainly know about it the next morning. New situations require the body to move and react in unaccustomed ways, and they always produce a degree of pain and resistance.

Think back to your teens, when you first fell in love. You experienced such a flood of painful and confusing sensations that you probably turned to love poetry or pop music in order to make sense of what was happening. With the help of a romantic song, it became possible to have your own feelings reflected back to you, and in this way, you were more able to deal with them. These feelings may have been uncomfortable at the time, but they were part of being alive, growing, and responding to new sensations. Rather than being avoided at all cost, pain, discomfort, and disorientation should be taken as evidence of change and a healthy response to life. A baby's first breath is a cry. We learn to walk by falling. We only cut our first teeth after experiencing a degree of discomfort in the mouth.

When I was a child in Liverpool, England, there was even a disorder called "growing pains," which was supposed to involve sudden twinges in the joints and were the price you paid for growing taller! Of course, there are painful feelings associated with growing up, leaving home, having our children marry, and experiencing the death of someone who has lived a long and fruitful life. The death of a parent can be painful, yet it is also a signal that one has entered a different phase of life and must assume new responsibilities.

Cézanne found his creative sensations "painful." Professional actors and musicians experience extreme discomfort before going on stage, ranging from feelings of panic to actual vomiting. At times, actors may even wonder why they put themselves through such torments nightly. But these uncomfortable feelings are indications that a professional has reached the peak of concentration just before walking on stage. If the feelings were ever missing, an actor or musician would worry that he or she would be unable to give a good performance.

Traditional societies recognize the pain associated with life's natural transitions and celebrate them through special rituals. These rites of passage are themselves often frightening or painful. This means that the normal feelings associated with growth are formalized and played out in a special ceremony, whereby they are acknowledged by the society as a whole. In this way, a rite of passage is attained speedily and with little lasting damage. A vestige of this can still be seen in our own society in the "hazing" and other initiation ceremonies for those entering a university or in the formal rituals associated with graduation and entry into a profession. Creativity on the part of the whole society enables individual transitions to take place in a healthy way.

Tragically, our own society has been unable to find a way to celebrate the coming of age of our young teens. Some sociologists have even suggested that young people have resorted to their own rites of passage in the form of body piercing, high-speed driving, participation in dangerous sports, and so on. These activities are, unfortunately, often associated with an excessive use of alcohol or drugs and other self-destructive behaviors. The pain of passing through the teens cannot be underestimated, yet, like the pain of childbirth, it appears to be something we adults too easily forget. We have become unable to take full responsibility for the youth who will be our own successors.

Not all pain is associated with the positive stages of growth and change. There is also the pain of guilt or, more often, the persistent, nagging edge of discomfort. Such feelings are associated with the realization that one has deeply wounded another or oneself. Guilt is said to "gnaw" and "eat away" at a person. It "stunts" her development and deprives her of the joy of life. At times, a person holds onto the pain of guilt as a form of self-punishment. Guilt can extend to the point where it becomes self-indulgence. As long as a person continues to blame and deprecate herself, there is no real motive for change, for breaking a destructive habit, or for moving on to a new phase of life. Moving beyond guilt requires a willingness to rescind pain in favor of a clear and honest acknowledgment of who one is and what one has done in the past.

The Catholic Church instituted the sacrament of confession to deal with guilt and the dangers inherent in hanging on to its pain. (Of course, it is equally true that some of this guilt was engendered by the proscriptions of that same church, particular those that are sexual in nature.) In confession, a person is provided with a formalized container in which to confront himself. In this way, he verbalizes the injuries he has caused and asks for forgiveness. Confession is a ritual for bringing things to an end, for facing oneself in honesty, and for being willing to close a chapter in one's life. In being forgiven, one may be asked to perform an act of penance. This is a symbolic acknowledgment that a degree of atonement is called for and must be completed. Confession and absolution free a person from the ties of time, they put the past in perspective, and they deal with the pain of guilt and sin. They acknowledge that being human means having the freedom to make choices and, at times, to make mistakes. Yet to continue through life with the knowledge of our imperfection requires a deep sense of the transcendent.

There are also injuries too intense to be tolerated and traumas occurring too early in life to be metabolized by body and mind. Moreover, there are individuals who are too sensitive to face the pain of living. When pain of this type cannot be dealt with, it becomes repressed and locked into the body and unconscious mind. When solidified pain gives rise to neurotic symptoms and patterns of self-damaging behavior, people often turn to psychotherapy.

A person in the grip of a neurosis may continue to repeat her self-destructive behavior. She may believe that she has escaped from one damaging relationship only to plunge into another with the same pattern. The aim of therapy is to break these self-limiting cycles and free the individual for creative growth. We have considered a number of approaches to therapy in this chapter. Let us now look more closely at one aspect of the famous talking cure first proposed by Freud and later modified by Jung and all the others who came after.

Such a therapy is only really effective when it works in an integrative way, so that thoughts, emotions, body feelings, memories, and so on cease to be fragmented and rigidified. People in therapy receive insights about the patterns of their life and the way these patterns connect to past trauma. After several years in therapy, a person will know considerably more about himself, but if this self-knowledge is not deeply connected with feelings and corresponding body sensations, it could not be said to have touched his life in any profound way. Like a smoker who always promises to give up smoking when he has finished the last cigarette in the present packet, there are those who claim they fully understand why they go on with a pattern of mistakes but feel incapable of making real change.

So at what level does real and lasting change take place within a therapeutic relationship? It is most effective, I believe, when the therapist is willing to be at risk. Under conditions of creativity, patient and therapist can resonate to such a profound extent that, to a certain degree, each experiences the body sensations, feelings, and memories of the other. Therapy often involves a patient revisiting situations of considerable pain and stress that occurred early in life. But what if these sensations can also be shared with the therapist and not simply at the verbal level alone? Therapists' training and maturity have better fitted them to contain and deal with such discomfort. In this way, it is possible that deep blocks to living can be discharged in a healthy way.

Psychotherapy has proposed a variety of mechanisms (along with corresponding technical terms) to explain such resonances. Some Jungians speak of "synchronicities"—meaningful coincidences of inner states (dreams, memories, thoughts, and fantasies) that have some correlation with external events yet are without any direct causal relationship. One therapist, for example, had a detailed dream about a woman who was identified as a Sioux (Native American). Later, he worked with a patient named Sue whose life paralleled the life of the Native American in his dream.

Another mechanism is called "projective identification." Here, an aspect of a patient's ego—memories and feelings— takes temporary residence in a therapist's mind. Yet another mechanism is "countertransference." In transference, the patient projects onto the therapist aspects of some key figure in her life (for example, a mother or father), along with all her unresolved feelings. In countertransference, the therapist experiences the feelings of the projected image directly. (Without a degree of countertransference, the therapist merely serves as a blank screen onto which the patient projects thoughts, feelings, and fantasies; no true resonance occurs.)

In *New Dimensions of Depth Analysis: A Study of Telepathy in Interpersonal Relationships* (1954), the psychoanalyst Jan

Ehrenwald gives examples of dreams reported by his patients that contain detailed events in the therapist's life that are unknown to the patients. Ehrenwald also refers to *enkinesis*, a term he gives to the sharing of somatic symptoms between patient and therapist. Likewise, the Jungian analyst Nathan Field, in *Breakdown and Breakthrough: Psychotherapy in a New Dimension* (1996), accepts that an aspect of the psyche of a patient can lodge itself within the psyche of the therapist.

For example, during a particularly unproductive session, a therapist may have a sudden, inexplicable desire to caress the genitals of her patient or, alternatively, to express a sudden burst of anger. Attending to these situations in a serious way and not translating them into action enables the therapist to develop the necessary insight, energy, and intensity to allow the patient to reveal some deeply hidden fantasy about his mother. Such incidents are more common than we think.

Over the past years and in my research for this book, I have talked with therapists from a variety of schools. Provided they feel sufficiently secure, most of them will report unusual and inexplicable occurrences during their professional lives. Such events are not confined to therapeutic encounters. It is simply that the ante is much higher in therapy than in normal daily life. But one only has to think back to those occasions when one has fallen in love or when a member of the family has been in great danger to realize how closely we can be linked, to the point of sharing thoughts and feelings.

These resonances between therapist and patient should not be seen as a form of pathology but rather as a creative movement toward health. The patient is suffering from the past. Memories, feelings, and sensations have been locked within the body-mind. Instead of process and movement, there is arrest and blockage. The resulting internal rigidity produces repetitive and destructive patterns of behavior. A small child who has experienced anger and violence at the hand of a parent may, in later life, become involved in a series of masochistic relationships or provoke pain and punishment from a partner. Therapists would suggest that such a person is attempting to evoke and control in adult life what was unendurable in infancy.

Whatever theories or mechanisms one invokes, these rigidities block the full flow of creativity in a person's relationships, work, and maturation. As the patient relates the history of her life, fantasies, and early memories and as she associates with the material of dreams, the therapist comes to understand the overall pattern of the neurosis. In turn, the therapist may reveal some of this structure to the patient. But as long as this results in no more than an intellectual understanding, it does not reach down into the feelings themselves—or unlock the words that are trapped in the body—which means that no true change will be possible.

But what if this material is projected out from the patient and into the therapist? Earlier in this chapter, I suggested that one's natural creativity is found deep within the body-mind, in the form of processes and sensations only dimly visible within awareness. For these deep sensations to become the raw material for thought and action, they must first be made manifest. They must be projected outward into the world, where they can be contemplated, reflected upon, manipulated by thought, and eventually reabsorbed and metabolized.

Previously in this chapter, we saw how inner feelings, sensations, and body orientations can be projected outward into the world as speech, gestures, dance, paintings, music, scientific theory, and what you will. What I am suggesting here is that in therapy or in a love relationship, a similar act of projection takes place, but this time, the projection is not onto stone, canvas, or language but into the body and mind of the Other. At this point, the therapist begins to experience uncomfortable feelings and reactions, along with unaccustomed ideas and strange memories. Training enables a therapist to recognize, contain, and work with these sensations in a fruitful way. For the patient, the internal, which previously lay outside awareness, has now been externalized and symbolized. It has moved to a new point of creative freedom, where it can be reabsorbed back into the his body and life.

The metaphor of chemical catalysis may help. Some chemical reactions, such as the reaction between oxygen and hydrogen gas or air and the natural gas in a stove, should occur spontaneously with a great release of energy. But for this to occur, the constituent molecules have to approach each other closely enough to engage. To do so, they need to have sufficient speed and energy to overcome the natural forces of repulsion. Individual molecules simply don't have sufficient energy to do this. And so, the two gases remain in a position of unstable equilibrium. Nothing happens until someone supplies the kick of energy needed to get the reaction going. Light a match and your stove ignites; strike a spark and a mixture of oxygen and hydrogen explodes. Under normal conditions, these gases want to react but remain blocked from doing so. However, if they are absorbed on the surface of a catalyst—such as platinum metal—they have more time to encounter each other. They can also borrow a little energy from the catalyst and rearrange themselves so as to react. Thus, with the help of a catalyst, a previously blocked reaction can now take place with ease. The resulting molecules then fly off, leaving the catalyst unchanged.

In an analogous way, the therapist acts as a catalyst, absorbing a series of fixed tensions and neuroses from the patient. Once lodged in the therapist's body and away from the enormous investment of energy the patient makes in maintaining his or her neurosis, the tensions and neuroses become available for reflection, to the point at which they can be freed to move and react. Finally, this freed material is reabsorbed back into the body and mind of the patient. At this point, thoughts, memories, emotions, feelings, and sensations become integrated in new ways. They are freed to move in a more open way within the patient's life.

Therapy is the formalized, theoretic counterpart of a natural movement to health on the part of all organisms. It goes beyond the bipolar encounter of patient and therapist into the whole of society, nature, and the environment. In this respect, just how far does mind extend? Earlier, we saw that David Bohm believed that because his body was composed of the matter of the universe, its deepest laws could be discovered through an inner journey. This suggests a dimension of mind that reaches beyond even the collective unconscious and down into the physical matter of the world.

Recall Cézanne's remark, quoted in Chapter 2, about his feelings when working before the motif. "The Landscape becomes reflective, human and thinks itself through me. I make it an object, let it project itself and endure within my painting. . . . I become the subjective consciousness of the landscape, and my painting becomes its objective consciousness." Given the artist's other remarks, it appears that this is no mere metaphor or figure of speech. It is not so much something he "believed," for beliefs depend upon faith. Rather, it is a practical account of the way he acted out his life as a painter. For Cézanne, nature was something truly alive, something that spoke directly to him. In turn, his work assisted nature in becoming explicit.

As for myself, I feel that I can also access this consciousness in front of Cézanne's canvas. It is almost as if the consciousness of nature had been adsorbed upon the surface of the canvas, transforming it and giving it a particular activity in which we, the viewers, can also participate. We all become nature's participators, for we are all charged with creativity. The way this creativity can flourish in the lives of each one of us will be the topic of the next and final chapter.

chapter 8

Conclusions

In voyaging through the previous chapters, we have seen how creativity is found in every corner of the cosmos, from the transformations of elementary particles to the artist's studio, from the big bang origin of matter to the opening of an eye. But this observation also brings us face-to-face with a paradox: Although creativity is all around us and is a part of our bodies, we sometimes feel dull and uncreative and complain of being trapped in a boring job or tied to a meaningless routine.

Why should this be? We want to feel creative. We admire creativity in others. Our society praises creativity to such an extent that it places individuals such as Mozart, Newton, Einstein, da Vinci, and Michelangelo on special pedestals with labels reading "genius." And yet, we find ourselves asking: Why can't we be like that? Why can't we produce one outstanding thing? Why do we feel so dull and uncreative? Why does life get us down?

Leaving aside illness, depression, and serious neurosis, frustrating feelings are mostly the result of the particular situations in which we are forced to live and work. In fact, the paradox of creativity is resolved by still another paradox. Our sense of dullness and of being trapped is not evidence of any lack of creativity—it is the direct opposite. It is a warning light telling us that our natural state *is* to be creative but that this creativity is not being fully engaged or challenged. If each of

us would only give serious attention to those feelings, then, together and in many subtle ways, we could bring about a
change in our society and the world in which we live.

People romanticize about the countryside only when they
have abandoned it for the city. As the singer Joni Mitchell put
it in "Big Yellow Taxi Cab," "You only get to miss what you've
got when it's gone. They paved paradise and put up a parking
lot." Most of the world's people never bother their heads about
being or not being creative. They are too busy meeting life's
challenges and doing what they do best. It is only those of us
caught up in all the mechanisms of the industrialized world
who have become obsessed with "being creative"—something
that we are going to be anyway, if we could only stop worrying
about it and get on with the business of living.

Let us look, for a moment, at the way people live in traditional and indigenous societies and how they have organized
themselves for generation upon generation. In many cases,
these societies function through small groups. The Plains Indians, for example, traditionally hunted buffalo and traveled in
groups. At night, they would sit in a circle around the fire and
pass around the pipe. During this ceremony, the person who
happened to be holding the pipe would speak from the heart,
directly to the others. The next in the circle might raise some
other issue; someone else would sing, offer a prayer, or tell a
joke. And so, the talk would flow around the circle and long
into the night. The questions, problems, and difficulties raised
did not really belong to one individual but were an expression
of the whole group. Sometimes, an issue would be illuminated
by an elder telling a traditional story to cast that particular
question or problem in a new light.

What was important in such societies was not so much the
conveying of specific information (although that, in itself,
could be important) but the general flow of meaning, a flow
that held the group together and in which everyone had an
obligation to participate. An important aspect of that same
flow was the way in which people walked around the camp or
village each day to exchange gossip. The significance of all this
can be judged by the fact that the most severe punishment in

this society was banishment—which implies being cut off from this flow of meaning.

In the eastern woodlands, the Iroquois do something similar from within the longhouse. And I remember seeing a television program from Africa in which herdsmen were debating whether to move on with their cattle to a new area of grazing. They sat all afternoon under the shade of a tree, talking, examining alternatives, retelling old stories, and calling on the elders of the tribe. In the end, they simply decided to move just a short distance away.

I don't wish to elevate such societies unnecessarily or to suggest that they are not without their own tensions and problems. But one thing is clear: Within them, everyone contributes to and draws from the society's general flow of meaning. The entire society is not so much a collection of persons but an expression of a creative flow that is ever changing, ever renewed. Within such a society, there is really no room for feeling uncreative or trapped. There may even be less of a sense of an ego or a self that is forever fixed, a self that must fit, like a part of a jigsaw puzzle, into job, family, and friends. Instead, the ego is somewhat fluid, a reflection of the way a person's name changes during his or her lifetime. Rather than the society being a collection of individuals who have come together for self-protection, the individual is more an expression of the group, just as a finger is an expression or function of the arm and hand.

Such societies meet their practical challenges through the operation of small groups that form spontaneously to carry out particular tasks. They may be created for hunting, gathering, clearing land, herding animals, fishing, building a longboat, moving a village, making a long voyage, or organizing a special ceremony. They come together to engage in a certain task, and when that task is ended, they dissolve back into the society in a natural way. Such groups are analogous to what chaos theorists call "self-organizing systems." So, although it is true that traditional societies are circumscribed by a variety of rules and taboos, they also have the ability to structure themselves in a variety of ways. (Self-organizing systems are widely found in nature. They occur when a system is open to its environment, not closed off or shielded. When energy, matter, or information flows through such a system, it will spontaneously organize to form a structure localized in space and time. Such structures can be quite stable; they preserve their own inner dynamics and can even repair themselves. But this rule applies only as long as they remain open to their surroundings. Once the flow-through of energy, matter, or information ceases, the systems die away.)

Such groups will probably not be hierarchical, although they will generally have a leader. But then, leadership within a self-organized group is something quite different from that in a rigidly hierarchical organization. In the former case, a person becomes a leader because of special skills, knowledge, and experience relevant to the task. It is not so much a case of leaders imposing their authority upon others; rather, their authority derives from the group itself. When the task is finished, that sanction for authority is withdrawn, the group dissolves, and the leader goes back to being an ordinary member of his or her society.

While the group is operating, each member has an obligation to work in harmony with the others and to use his own special skills for the good of the entire group. Working in this way places demands upon a person's creativity but, at the same time, rewards him with a feeling of common purpose and shared meaning.

Now contrast this with the environment in which most of us work today. If life seems dull, it is because we are not participators in a common, shared pool of meaning and because our creativity is not being engaged to the full. Our organizations are often rigid and strongly hierarchical. Leaders exert authority upon employees without the employees ever sanctioning managers and bosses to have such power. In most cases, the origin of power comes from outside the organization, from the shareholders or board of directors that has elected a chairperson and officers. Thus, the source of authority is extraneous to the people who are actually doing the work.

The organization itself may have been established for any one of a variety of reasons—to sell goods, fill up a space in the marketplace, offer a service, or fulfill duties as a branch of government. It will have a mission or goal, which may have been explicitly stated decades earlier when social conditions or the marketplace was different. Moreover, the organization will have a relatively fixed internal structure of authority and control.

Certainly, management consultants have become aware of the new scientific metaphors, including chaos theory and its notions of self-organization and open systems. This is having an effect on the way that modern organizations are run. But in most cases, organizational structures are still modeled upon the pyramid. In turn, this geometrical structure is reflected in its various lines of communication and even entrenched in the architecture of the building itself—which, in many cases, is an office town with the chairperson and managers on the top floors and lowly workers in open-plan offices below.

The result is that most people in an organization are not fully engaged. They are not sharing in and contributing to a common pool of meaning, which should be the true life of the organization. Remember Chapter 5's discussion of "mental spaces," those spaces that arise when two people are in active communication and giving attention to each other's words, gestures, and tone of voice. Each person is busy creating and furnishing an internal space, triggered by the conversation; in turn, as the conversation continues, that private mental space is transformed into a shared field of meaning. In the same chapter, we also learned about the way our brains respond to this ever changing pool of meaning and modify their structure in subtle ways.

This is how we should be engaging every dimension of our lives. But unfortunately, most businesses and organizations are limited, relatively mechanical, and structured in such a way that they do not involve us fully. And so, the organization is populated with highly creative people yet is only capable of engaging the potentiality they represent in limited ways. We feel dull because we sense that we are never being called on to contribute what we could do best. We watch the clock and wait for that moment when we can leave work and put all our creative energies into family, friends, or hobbies. How enlightened are those organizations that allow employees to take off and work at home if they want. With a personal computer and a modem, people can work anywhere. Writing book after book from a tiny Italian village, I can dialogue with people all over the world, carry out research, and even coauthor a book with someone in another country. We get the most out of people when we allow them to work at their own pace and in their own way upon a common task that truly engages them.

Huw Weldon, head of Television Arts during the BBC's most creative period in the 1960s, was once asked how he ran the arts department. Did he give the public what they wanted? Or did he attempt to educate them by including programs on Mozart or contemporary painting? Weldon replied that he took neither route but simply hired the most talented writers and directors he could find and then let them do exactly what they wanted. The results were some of the best documentaries and arts programs that the BBC ever made.

Under such conditions, when authority is given back to individual employees and when the field of meaning that is the organization flows from each person, everyone feels engaged and is using his or her creativity to the full. Our feelings of being trapped and constrained are warning bells telling us that this is not happening. The result is that we become like plants that are suffocating because they are being kept from the air and sunlight.

Whether boards and managers like it or not, this state of affairs isn't going to continue much longer. The future has arrived. Already, people are creating virtual organizations and virtual corporations. Very probably, virtual universities have already arrived. Thanks to modern communications, it is possible for groups of people to spontaneously self-organize in a variety of ways and to transform again into some other structure when the environment changes. A person need not work for one organization for the rest of his or her life. In fact, such a person could be engaged by several virtual organizations at the same time and working on a variety of tasks that complement one another.

This is the wave of the future. Work will be not only more effective but also much more rewarding for us. It will involve us to the full and draw upon all our inherent creativity. Organizations of the future are going to be more effective and far more creative than anything structured along traditional lines. They will be like a new, cooperative species that has suddenly made its appearance and already has its own built-in ecological niche.

But here, I must add that working at home with a computer and modem has its limits. While one is totally engaged in virtual worlds, the other, social side of existence is being neglected. It is important to meet people face-to-face and to shake their hands, share a meal, walk together, and discuss the trivia of life. Virtual organizations are going to work best when they also give time and attention to the physical needs of the nonvirtual—the body as well as the mind. But this may also happen. The more people work at home or in small, self-organized groups, the less need they will have to commute to business in a crowded city. Cities will certainly remain, but their functions may change; with their art galleries, opera houses, theaters, parks, historic buildings, and so on, they will primarily be places of recreation. As people work in more spontaneous ways, we could well see the rise of self-governed small communities in which more attention is given to the environment, civil architecture, and general quality of life.

Energy

When we feel dull and uncreative, we also tend to get easily exhausted. We may be eager to get home from work only to find that the pleasure of seeing our children or working in the garden becomes a chore. We are drained and unable to set aside what energy remains for tasks we would normally enjoy.

Why should this be when energy is flowing all around us through the universe? Nature's systems are powered by sunlight, water, and air in abundance. Rivers, waterfalls, weather fronts, patterns of sand in the desert, and countless other structures are all maintained by the ready flow of energy and matter. In Chapter 4, we found that the quantum vacuum state of the universe is an inexhaustible source of energy that manifests itself in the endless transformations of the elementary particles.

The universe is freely giving out energy because it wants to sing for joy. We and the birds, flowers, and stars all express ourselves in this song. Energy is readily available to us. We may take as much as we want. With it, we create, and in creating, we pay back this energy. Actors and musicians have told me how much they gain from an audience. An actor may have spent the long day rehearsing the next play in the repertoire. A musician may turn up at the concert hall after an exhausting journey or on yet another night of a nationwide tour. They summon up what energy they have left and make their entrance on stage. And then, as they begin to play or act (and provided that the audience is with them), they experience an enormous surge of energy. The audience is freely giving to them, inspiring them, and helping them to find new dimensions in their work.

I once talked to the pianist Alan Rawlinson about the problems he had experienced in interpreting one of Beethoven's late piano sonatas. By the end of his life, Beethoven had reached such a dimension of musical maturity that he could almost write in shorthand, taking dangerous leaps and following unexpected developments. Playing such music requires a great deal of study to discover what Beethoven meant and find a way to convey to the audience what is happening in the score. One passage resisted all of Rawlinson's efforts in this regard. He compared the experience to coming across a passage of writing with all the punctuation marks left out.

Finally, when playing at the Wigmore Hall in London, he realized that the audience was totally silent and exhibiting an intense degree of concentration. "As I came to that place, I knew what it meant for the first time," Rawlinson said, "the grammar of it became clear. I felt it was a miracle, a stroke of genius on Beethoven's part. It could not be any other way! Only afterwards could I explain it intellectually—how he could make such a jump under certain conditions. It was like a ray of sun coming into the music." Rawlinson went on to say that his insight into the meaning of the music "could not happen unless all those people were contributing." He also added that in that moment, "the body was doing its job so well that it was almost like not having a body" (Rawlinson, in conversation with the author).

Such are the gifts shared by actors, musicians, and athletes. They give off pure energy to the audience or spectators, and in a way that appears to defy the physicist's laws of conservation. The audience is giving energy to the artist, and the artist is radiating energy to the audience to the point at which the sum of this energy vastly exceeds what was there to begin with.

When we are engaged, healthy, and creative, we are conduits for unlimited creative energy. But if such energy is always available to us, why should we ever feel drained? The answer is that, rather than lacking energy, we are actually diverting it into uncreative paths. At times, our fears about life, the painful memories and traumas we hold inside ourselves, and our reluctance to change and embrace new situations cause us to become rigid. We resist movement and block change. All this acts to divert our naturally flowing energy into maintaining a fixed structure against the natural processes of growth and transformation.

We are like one of those organizations we met earlier in this chapter. The social and economic conditions under which the organization was created have changed. Yet, rather than transforming its whole structure, the organization continues on in the same rigid way and drains its assets as well as its employees' time, just to remain in business.

To overcome this dilemma, we must become reconnected and engaged. If we can but regain our place in the natural flow of things, we need no longer support all these rigidities, and we will thereby free our natural energy for more creative work. Let us therefore look a little closer at the important issue of engagement.

One of the endearing but sometimes irritating things about young children is the way they will ask you to read the same story over and over again. It's something of a ritual because the reading must always be repeated in exactly the same way and with the same voices used for all the different characters. We wonder why the children don't get bored and turn the page to get on to something new. But no, they want to go back and hear the same story all over again.

Does this mean that the children are dull and uninterested in new things? It does not, for within that story, they have found an inexhaustible treasure to mine. They are visiting it over and over and making sense of the thoughts and feelings that are flooding through mind and body. There may be a variety of reasons for choosing that particular story. Something about it may strike a chord. Maybe it deals with a puzzle about the adult world, touches on some basic fear, or addresses the issue of separation. The child psychologist Bruno Bettleheim argued that a child first touches upon the deep archetypal issues of life through fairy tales, which is why they are revisited so frequently.

Whatever the reason, each time we repeat the story, it acts as a matrix in which the children can find their bearings, negotiate, and explore. If they were always to ask for a different story, they would merely be seeking novelty. But listening to the same story over and over again indicates the great creative engagement active in even the youngest child. A youngster may also be captivated by a favorite toy or a repetitive game, such as putting stones in a bucket and then taking them out again. In all cases, the preoccupation is intense.

While I was writing this chapter, I saw a photograph that perfectly sums up this point. At the time, the Kosovo crisis was at its peak, with the bombing of Serbia in its eighth week and tens of thousands of refugees flooding into neighboring countries. One morning, the newspaper carried the photo of two-year-old Bosnik Tusha, who was staying in the Macedonian refugee camp at Stenkovec. Like others who had fled from the massacres, he and his family had no possessions, and the related headline suggested the tragedy of a child without toys. But, in fact, the photograph showed a young boy totally engaged; he was wearing his father's shoes and pushing a cardboard box along the earth. Maybe, for a time, the child had become his father driving a truck. Or perhaps he was playing out the trauma of a refugee's flight—who knows? But I couldn't help thinking that in the midst of all that tragedy, with entire villages burned down and male adults massacred, a child was using minimal materials and his natural sense of play to create a world of his own (*The Observer*, May 16, 1999).

Beyond those executive toys on our desks, we adults have ceased to play and are unlikely to lose ourselves in the task of filling a toy bucket with stones. But in other ways, we can still live in the world of the imagination. This happens when we merge horizons with a book. Think what happens when you turn to a favorite poem or reread a well-loved novel. The work is something you trust because it is familiar, comfortable, and well visited. Yet each time you visit it, it is both the same and yet totally different. We can engage the work in a multiplicity of ways. We reread it with total engagement because we know we can always find something thoroughly new within it.

Of course, the words on the page are exactly the same. The difference lies in the many dimensions in which we can engage the work and the creative effort we can bring to it—seeking allusions, unpeeling a new level to a metaphor, and making fresh associations with the world outside the written words. As we are engaged in the act of reading, we are creating the poem or book anew. The author urgently needs us, for without our input, the work is dead. It can only live to the degree to which we are prepared to engage and animate its form.

Think of a theater director who is asked to put on *Hamlet* or *Macbeth*, two plays that have been produced repeatedly in countries across the world. What on earth can the director do with the play so that it will engage a sophisticated audience who have seen productions before? How can the audience be persuaded that the play is not a museum piece, that it is almost freshly minted and speaks of something totally relevant to their contemporary condition?

The director must be aware not only of the history of Shakespearean productions but also, in a modern setting, of the whole of world theater. Add to this an awareness of the social and political tenor of the times. Armed with all this, the director returns to the text for a period of intensive study. Sometimes, this preparatory work is done in private; at others, it is done with a company of actors who discuss the text, try out various readings, and explore scenes and relationships through improvisation. When the director and actors are creative and talented, something begins to grow and the play takes on a freshness and displays angles that may not have been seen before.

Let me take one example: Peter Brook's production of *A Midsummer Night's Dream*, which I saw in London in 1971. The play can be something of a warhorse when put on by tired repertory actors for school visits. Or it can be a lavish production, complete with forest sets and courtroom scenes. Generally, it is played for its enchantment, for it is a drama about fairies with a dose of comic business thrown in to get a few laughs as the "rude mechanicals" put on their play within a play. But when Brook, a director of vision, came to work with England's Royal Shakespeare Company, he dispensed with lavish sets and mounted the play in a white cube. The actors were confined within a space. The play was focused. The heat grew, and we witnessed an intense exploration of the nature of sexual jealousy.

Today, Brook's *Dream* has entered theater history. Although it has been filmed, the film itself is not a piece of theater, and so the performance lives only in the active memories of those who saw it. Future directors cannot afford to ignore it. They must take Brook's particular reading into account as they go into rehearsal and seek the new within Shakespeare's text.

What Brook and his actors did for Shakespeare, what a small child does with a fairy story, or what we do when we reread a familiar poem can also be applied at every moment of the day to a relationship, the environment in which we live, and even an entire life. We can use our energy and our creativity to merge horizons with the world, to engage it directly, and to re-animate and renew all that lies around us.

Creativity can be expressed in a number of ways. Take, for example, a friendship. Friendship brings with it obligations. It needs renewal. It is something that must be revisited from time to time. Old contacts must be taken up again in letters, or phone calls must be exchanged. Creativity can also be exercised in gardening or cooking. It can involve the decorating of a house, working in the community, coaching a baseball team, or putting on a dinner for friends. To be truly creative, you need not compose a symphony, paint a fresco, or write a novel. Simply giving attention to your life may engage creativity on a full-time basis.

All this may sound a little flaky to the cynical reader, so let's look at what contemporary artists are doing. There is a movement within the art community that speaks of "the death of painting." Of course, not everyone agrees with that assessment, and quite a few artists continue to paint, make prints, and carve stone. But others feel that neither the art gallery nor the collector's home is the primary place for their work. They have begun to think about the space outside their studio—the street, the city, the country, and society at large. How can they, as artists, relate to the world in which they live and address the issues of its changing values?

The German artist Joseph Beuys argued that "everyone is an artist," a statement that places an obligation on each one of us. We should not leave the making of art to a chosen few who shut themselves away in studios and workshops. Each one of us can engage in what Beuys called "social sculpture." The work of art then becomes focused not on canvas or stone but on social gestures and on working within the world.

But even here, a word of warning is called for. In giving lectures and going to conferences, I've met quite a few people who are busy solving the world's problems. Surprisingly, a few of them are living lives that are in quite a mess. They may have been involved in a series of dysfunctional relationships, be at odds with others in the same field, or be engaged in battles with foundations and institutions. (This reminds me of the adage that the shoemaker's children are the worse shod.) Sometimes, when your own life is in a shambles and its problems appear insurmountable, it is far easier to look outside, away from the problems, than it is to look within. And so, all the confusion and internal conflicts are projected outward onto the world.

People in such situations engage the world's problems with great passion, precisely because they see those problems as a reflection of their own, internal lack of harmony. And thus, they set about trying to resolve what lies outside. In doing so, they may gain insights and work out some of the conflicts that lie within. Yet how much better it would be to reserve some of that creative energy for one's own life and then operate from a more secure platform.

Being creative means giving great attention to one's whole being, to what one does and how one does it. It means taking nothing for granted and trying to see what may lie behind all the habits of body and mind. The performance artist Marina Abramovic gives intensive workshops for young artists. One of her exercises is to have total awareness of the act of walking, in part by slowing down the action to the point at which every shift of balance and every muscular movement becomes a conscious and intentional act. At the end of a week or more of intense work, a young artist's performance may consist of no more and no less than walking down a set of stairs—an action that could take half an hour to perform. This is the art of oriental theater, in which every gesture and every breath is part of the work and not accidental to it. It is present in the Zen maxim, "When walking walk. When sitting sit."

On a personal note, I once saw a beautiful Zenlike gesture made by the pianist Thelonius Monk. Monk, along with Charlie Parker, Dizzy Gillespie, Bud Powell, and Charlie Mingus, was one of the creators of bop. I saw him in Toronto during the late 1960s. By then, he was a legend, but the audience members in the club that night were of a younger generation and not as attentive as the pianist would have wished. At one point in a long piece with the combo, the other instrumentalists stepped back to give Monk room for his solo. He struck a chord, meditated a while, struck another chord, and then took out his lighter and, in a beautiful physical gesture that perfectly complemented the drum and bass accompaniment, lit his cigarette. He sat for a moment, struck a couple more chords, and ended what was a masterpiece of minimal gesture.

If you like, you can give all your attention to creating a scientific theory, writing, painting, or composing. But you can also create the curve and story of your life. Imagine yourself at the age of eighty. Will you be frittering away your last years in a retirement home? Or will you have sculpted yourself into a great historical monument? Have you given as much attention to this life as Michelangelo gave to a stone or Beethoven to a symphony? What will the pattern of its enfoldment be? What stories will your grandchildren tell about you? What music of life will you have left behind you? What change will you have wrought upon the world?

Gary Snyder, along with Jack Kerouac, Allen Ginsberg, Gregory Corso, and William Burroughs, was part of the beatnik movement of the 1940s and 1950s. Snyder felt that his group had changed the sensibilities of the American people (although the beatnik influence on literature may not have been as significant as it appeared at the time). His assessment was modest but accurate: "We moved the world a millionth of an inch" (quoted in Jon Halper, ed., *Gary Snyder: Dimensions of a Life*, 1991). That is a pretty impressive achievement. To have changed the world, even if only by a millionth of an inch, is a significant thing to have done. What if each of us could move the world just a thousand millionth of an inch? After all, there are an awful lot of us, and the total shift would be perceptible.

Snyder's story has a sequel that shows how creative and far-reaching even the most casual encounter or gesture of life can be. Snyder appeared under the guise of Jappy Ryder in Kerouac's novel *Dharma Bums*, and like its predecessor, *On the Road*, the book was something of a bible to many young Americans, including one young man named Dan, who was working for the RAND Corporation. At the end of the book, Jappy, like Snyder himself, takes off to Japan to become a Buddhist

monk; coincidentally, in 1961, Dan's work (which involved analyzing the Pentagon's defense plans and nuclear strategies) also
took him to Japan. Recalling Kerouac's description of the famous rock garden in Kyoto, Dan decided to take a day off and
travel from Tokyo to the Zen monastery. Later, walking back to
his Kyoto hotel, he dropped into a bar for a beer, where he was
helped with the menu by two Americans—one a bulky former
Hell's Angel and the other a smaller, soft-spoken man who
turned out to be Gary Snyder. "You're the reason I'm in Kyoto,"
Dan told him, and the two men spent the rest of the day and
much of the night talking at Snyder's home.

Theirs was an unlikely combination, the aesthetic monk and
an expensively dressed representative of the American military
complex, but the two men seemed to hit it off and argued for
hours about the nature of pacifism. Dan came away almost—
but not quite—convinced.

In the years that followed, Dan's work took him to Vietnam,
where he became more and more uneasy with what he was
learning about American policy. By now, he was feeling the
weight of the defense secrets he was carrying, and in the end,
he decided that he had to make his knowledge public. So Dan
started to photocopy the Pentagon's top secret documents. His
intention was to offer them to the Senate Foreign Relations
Committee, even if disclosing the sensitive material brought
him a lifetime prison sentence.

Before he made the papers public, Dan decided to say good-
bye to a few of his closest friends. This small group included
Gary Snyder, for their one meeting had had a powerful influence on him. And so, with the as yet unreleased Pentagon Papers in the trunk of his car, Dan Ellsberg drove from Big Sur
toward Nevada City in search of Snyder's farm. In this way, the
two men had their second meeting. After the papers were released, Ellsberg was indicted and faced a maximum sentence
of 115 years. In 1973, the case against him was dismissed because of the U.S. government's misconduct.

In this instance, perhaps America itself was moved slightly
more than a millionth of an inch. Seeking material to discredit
Ellsberg, members of the White House's "dirty tricks" depart-

ment broke into the office of his psychiatrist; this was the same group of men who were bugging the Watergate Hotel in Washington. Ultimately, these events led to the move to impeach Nixon and his subsequent resignation. Thus, one chance meeting in a Kyoto bar was a key element in a long chain of events that rocked a whole nation.

Being aware of and making use of chance has always been a vital aspect of creativity. Artists rely upon chance as an important element within their work. Chance allows something Other to enter, something that is unplanned. Chance is the tip of in iceberg, a door into the infinite.

The painter Gainsborough splashed ink marks over some of his drawings to give them a special visual texture. Jackson Pollock laid his canvases on the floor and walked around them as he poured and dripped paint. After Marcel Duchamp finished his *Great Glass (The Bridge Stripped Bare by Her Bachelors, Even)*, it cracked in transport. Duchamp looked at the pattern of cracks spread across the work and decided that they added to and, in a sense, completed the work. He also preserved the dust that settled on the glass while it was in his studio.

In the end, however, it is not chance itself that is important but rather our attitude toward it. Accepting chance means remaining in a state of openness to the moment. That is where creativity lies. The theater director Peter Brook writes of the skilled actor who, over many days of rehearsal, builds up a performance. Piece by piece, a character is created and stage business is perfected. The skilled actor aims to hit the mark on opening night. By contrast, the highly creative actor is never satisfied with a performance. This is the actor who knows that no matter how much work is done, there always remains an elusive element of truth. Such an actor is bold enough to be willing to throw everything away on the first night and walk onto the stage naked. With such an approach, it is always possible that a truly great performance will be born.

Dan Ellsberg's chance meeting with Gary Snyder in a Kyoto bar is but one example of the way our actions can ripple though society in totally unpredictable ways. A chance remark made by a teacher may spark a lifelong interest in a student who is primed for that very moment. Individuals who make a tiny stand on a matter of principle can suddenly find a whole nation behind them. But these are exceptional examples. In most cases, the best we can do is remain aware and alert to every situation and try to act kindly and with an inner sense of authenticity. We must trust in the natural world, trust that authenticity, kindness, and love will move though a system and act to renew and sustain it.

However, there may be times when we feel that this simply isn't enough, times when we know that things are wrong and we must do something about them, times when we are up against a system that is corrupt or so rigid and unresponsive that it is damaging people. Although action is needed in such situations, our responses are likely to produce what could be called a violent action. It may not be violent in the sense of an act of warfare or armed aggression, but it could be an action that does violence to the fields of meaning shared by those who live and work in such situations.

Actions of this type are violent because they are reactive. They come from our internal sense of anger and outrage. They are also violent because they are imposed on a system from the outside and do not arise in a natural way from the inner dynamics of the system itself. We look at a system and see it in terms of errors, defects, and deviations from what we consider to be its well-ordered operation. Problems are viewed as being localized at some point in the system. And so, we attack the problem at that region in which we perceive the defect, rather than at its source.

An image may be illustrative. The effect of such an action is rather like what occurs when someone throws a stone into a pond. The source of the action comes from outside the pond. It is not part of the pond's inner world. Such an action is violent at the point of entry and ripples off throughout the whole pond.

Now imagine a radically different sort of action, one that begins at the very edges of the pond as an almost imperceptible ripple. Imagine this moving inward, gaining strength by meeting all the other ripples that are simultaneously moving inward. Finally, these ripples meet in the center of the pond, where they produce a large disturbance. As a visual image, this is just the time-reversed picture of our first example. But as a metaphor, it is somewhat different. It suggests an action that is not extraneous but that begins within the system itself. Its origin is something very gentle and almost imperceptible but also highly sensitive and intelligent because it takes into account the entire dynamics and all the web of connections of the system. In one sense, it is an action that arises out of the system's entire field of meaning.

I have called such action "gentle action" and suggested that it arises when we are in harmony with the dynamics and behavior of the whole system, when we resonate with it and feel it in our bones. Rather than being an action that does violence to a system that is extraneous to it or an action promoted by some conceptual image of the system, it is an action motivated by kindness and compassion.

In a deeper sense, the source of this action lies within our selves, for when we work within an organization, live in a society, or participate in a family, we carry in our minds and bodies a representation of the structure and meaning of that family, society, or business. If we are angry and frustrated, then we are, in part, annoyed with ourselves. But anger, rage, and frustration can so easily distort and dislocate the action from clear perception.

This is why I have suggested that the origin of gentle action should lie in the suspension of action. In certain situations, we have a natural and immediate desire to act. Yet if the source of our action is in some way distorted, we may only end up making things worse. But how can we know when to act and when not to? First, we should remain just at the edge of an action, sensing it within the body—or, for that matter, within the organization. This is a little like holding a tuning fork near an instrument: When the instrument is properly tuned, the fork will begin to vibrate in sympathy.

If we suspend action, if we hold onto it but don't yet release it, we have an immediate awareness of all its implications and how it is going to unfold in time. We can detect areas of distortion within us and move to their source. Holding onto the edge of action is similar to the examples of creative holding we discussed in terms of an artist such as Anish Kapoor or a composer such as Michael Tippet. In each case, the intention to act, to manifest within the outer world, must be held inwardly until it has come to perfect fruition and is ready to be projected outward. Creative suspension may take days, hours, or a mere fraction of a second. What matters is acting at the right moment.

The example I have given involves an individual seeking to act within society. But creative suspension and gentle action could equally well apply to a group of people who are seeking to act together. Native Americans, for instance, tend to act only "when the time is right," a response that is a particular mystery to non-Natives. One day, a group of Native Americans announce that they are going to hold a ceremony or go hunting. Then nothing happens. The morning drags on, and no one appears. And so it goes on the following day. Then, at dawn one day, a group begins to congregate to carry out a certain task. They assemble quite spontaneously, without anyone calling them or acting as leader. The need for action was always present, but it was contained by the group and not set into motion until the time was exactly right and the action perfect.

Conclusions

As we have seen, creativity is ubiquitous and entirely natural. It may begin outside the light of conscious awareness until it is ready to manifest itself in one of a wide variety of ways. For such reasons, creativity cannot be legislated or planned. It cannot be commanded to appear. Neither can it be taught.

Musicians, painters, scientists, actors, chefs, and others can be taught various skills and must serve long apprenticeships before being able to practice their crafts. But none of this is "teaching creativity." Similarly, all of us can learn techniques to disrupt persistent habits of thought and free us for new ways of thinking. Brainstorming is one such example.

Most of us have a desire to protect our egos and have others think well of us. The natural consequence is to try to hold onto our ideas and push other people to accept them. But this is hardly the way to achieve true cooperation. In brainstorming, everyone throws out ideas that may be written on large sheets of paper. Once set down, the ideas are the property of the group as a whole and do not belong to a particular individual. Then follow periods in which these ideas are elaborated, with all their good points being emphasized. Next, they are criticized, with their drawbacks being weighed. In this way, the group as a whole engages in a creative enterprise and moves beyond the sort of blocks that often crop up in a working group.

Another approach that some people find freeing and helpful is Eduard de Bono's "lateral thinking," in which you learn to come at questions from surprisingly different angles so that what previously appeared to be an insurmountable problem simply vanishes.

And here is something else you could try: If you have an art gallery near you, don't wander from room to room the next time you visit. Instead, pick a single painting and see just how long you can stand in front of it and be actively engaged. Maybe you can engage the painting in a conversation, a two-way flow of questions and answers. The more you ask of the painting, the more it will give back to you, and, as you become engaged, you will be animating the work and giving it new life in return.

You could, for example, begin to ask about the way the painter's society differed from your own. How would his or her contemporaries have looked at this painting? What was their knowledge of the world? How is it that the painter still manages to speak directly to you across the centuries? Some artists (and they are not all painters) tell me that, in their work, they feel they are engaged in a dialogue with the whole history of art. Does this make sense to you as you stand in front of a painting?

Or you may be struck by the color in the painting. Is it used in a symbolic or heraldic way? Is it "naturalistic"? If so, then ask whether things in nature really look that way and why they are depicted in such a fashion in the painting. How does the color contribute to a sense of depth and space? Is it being used for its emotional impact? If so, how? And how are the colors in the painting working together?

Already, though some time will have passed, you are still engaged with the painting. You begin to question its forms, the use of line, and the way its overall structure has been engineered. Or you might notice how it evokes paintings from quite different historical periods and movements. Maybe now, the various questions you have been asking are starting to come together. Maybe they are sparking others.

Now could be the time to go home and get out a book about that artist and his or her period. Then go back to the gallery and engage that painting yet again. Come to know it like a friend, and always be able to dialogue with it. For me, there are a few paintings like that in galleries in New York, Chicago, Boston, Paris, London, Florence, and Siena. When living in London, for example, I used to visit Piero della Francesca's *Baptism of Christ* and Seurat's *Bathers, Asnieres* on a fairly regular basis (along with quite a few others), and these works and I still continue to dialogue together as good friends.

Now that you've found that you can engage a single painting (or poem or piece of music) for longer and longer periods, why not see if the same process can be applied to the natural world? Simply choose a place and sit. Don't try to do anything or to impose any order on what you see. Just sit and discover what the environment wants to say to you. Native American elders have told me that the natural setting was their grade school and university. They sat and watched a trail of ants, a bird, or a small animal hour after hour, day after day, until they began to know its world and understand its inner nature.

Maybe you will also find you can do this with a friend or partner. But even more important is to hold this engagement with yourself and your whole body. Simply be with yourself. Don't impose some sort of life plan or concept about yourself. Don't judge yourself. Don't fantasize about what you could be.

Simply be with yourself, just as you engaged that painting for longer and longer periods.

There is no goal to such an exercise. It's not about self-improvement or accumulating knowledge and wisdom. It is simply an act of engagement with one aspect of the world that should be most familiar to you—your self. In the end, approaches like this one are no more than simple techniques to get us to think, give attention to the world around us, and do things in different ways. They are not creativity itself. They are simply means designed to remove some of the blocks to creativity's free flow.

Creativity is ever present. It streams through the firmament. It envelops and nurtures each moment of time, and, if we will but allow it to enter, it will flow through our lives and embrace us. But, like any gift freely given, it will also foster an obligation within us, for gifts should be exchanged. And so, we must take that creativity that is offered to us, in whatever form it comes, and try to use it wisely. We must use it with kindness, compassion, and engagement. In this way, its fruits will grow and multiply, and so we can return them to society, nature, and the cosmos. The rest is up to us, for with intelligence and love, we too can move the world by that millionth of an inch.

Index